FORAGE FOR KANSAS

A MANUAL FOR SEED RETAILERS AND FORAGE PRODUCERS

By: Stephen Ahring

NOTICE TO BUYER

The information and instructions contained in this publication are believed to be reliable. However, the author and his employer DeLange Seed House, Inc, do not guarantee the accuracy, completeness, or applicability to local conditions of the information and instructions herein. They assume no responsibility for errors or omissions and disclaim all legal responsibility.

Likewise, any trademarks, herbicide brand names, varieties, or any other intellectual property owned by another party and referred to in this manuscript are only used for reference. The author is not endorsing specific products.

ACKNOWLEDGMENTS

Nobody ever writes in a vacuum. Without the support of friends and family, little of importance is accomplished. This is especially true in my case. Jill DeLange took my handwritten scribbles and transformed them into this manuscript. My friend, Terry Johnson spent many hours editing the text. My good friends and fellow seedsmen, Tom Lutgen of Star Seed and Maurice Miller of Sharp Brothers Seed Company, reviewed the manuscript and made suggestions that improved the content.

And lastly, I need to acknowledge the tremendous work of our Land Grant Universities. The agricultural extension services of these universities provide an invaluable resource to producers and agribusinesses.

DEDICATION

This book is dedicated to Howard and Carla DeLange. As a graduate of Kansas State University in May of 1983, my prospects for employment seemed bleak. The USDA had implemented "Payment in Kind" (the PIK Program) in 1982. As a result, large chunks of farm ground were being idled or set-aside. Agribusinesses, such as grain elevators, fertilizer dealers, implement dealers, agrochemical dealers, and seed companies were all very concerned about these reductions in crop production and the fact that farmers would not need to purchase as many inputs or services. They simply were not hiring.

When I interviewed with DeLange Seed, a whole new aspect of production agriculture presented itself. DeLange Seed was struggling to keep up with the demand for cover crop and soil improvement crop seeds to plant on these set-aside acres. While large multinational agribusinesses were seeing nothing but lemons, DeLange Seed was busy making lemonade. I was impressed and in the thirty years since, I have never regretted accepting the position of Company Agronomist for DeLange Seed.

Howard and Carla DeLange not only provided the opportunity to work in the industry I love, they also taught me the seed business and encouraged me to take an active role in industry associations. For this, I will be forever grateful.

TABLE OF CONTENTS

	Pages
Introduction	i - iii
CHAPTER ONE: NATIVE GRASSES	**1–29**
Little Bluestem	5–6
Big Bluestem	7–9
Sand Bluestem	10–11
Switchgrass	12–13
Indiangrass	14–15
Sideoats Grama	16–17
Blue Grama	18–19
Buffalograss	20–21
Eastern Gamagrass	22–25
Wheatgrasses	26–29
CHAPTER TWO: INTRODUCED GRASS	**30-56**
Smooth Bromegrass	33–35
Tall Fescue	36–39
Orchardgrass	40–41
Ryegrasses	42–44
Timothy	45–46
Reed Canarygrass	47–48
Kentucky Bluegrass	49–50
Bermudagrass	51–53
Crabgrass	54–56

CHAPTER THREE: WARM-SEASON ANNUALS

	Pages
	57–79
Teff Grass	59–61
Sudangrass	62–63
Hybrid Sorghum/Sudangrass	64–66
Forage Sorghum	67–68
Hybrid Pearl Millet	69–70
Foxtail Millet	71–73
Corn & Sorghum Silage	74–75
Prussic Acid	76–77
Nitrate Poisoning	78–79
CHAPTER FOUR: SMALL GRAINS	**80–98**
Winter Wheat	87–89
Oats	90–92
Barley	93–94
Winter Rye	95–96
Winter Triticale	97–98
CHAPTER FIVE: LEGUMES	**99–137**
Alfalfa	101–105
Red Clover	106–109
Crimson Clover	110–111
Alsike Clover	112–113
Arrowleaf Clover	114–115

	Pages
Sweet Clover	116–118
Korean Lespedeza	119–121
Ladino Clover	122–125
Birdsfoot Trefoil	126–129
Crown Vetch	130–131
Hairy Vetch	132–134
Winter Field Peas	135–137
CHAPTER SIX: BRASSICAS	**138–145**
Turnips	140–143
Daikon Radish	144–145
Quick Seeding Chart	146–148
Glossary	149–153

Introduction

Forage for Kansas is not referenced, as the original intent was to use it 'in-house' as a manual for DeLange Seed's sales representatives and dealer network. However, farmers and ranchers who have seen copies have requested that it be made available. None of the material contained in this publication is proprietary. It is public knowledge and represents a common-sense approach to forage management systems. Much of the material was gleaned from old college textbooks and various university extension publications – all of which were filtered through 30 years of experience of the author in his role as Company Agronomist for DeLange Seed House, Inc.

The science of grassland management and forage production is a complex and challenging subject. Matching the nutritional needs of various classes of livestock to the seasonal distribution of yield and quality of forages is a daunting task. Considering that Kansas is geographically located in the center of the United States and serves as a transition zone from north to south and east to west, there are myriad choices of forage species that claim some adaptation to the Sunflower State and this compounds the complexity.

Forage for Kansas was written to help seed retailers and forage producers gain a basic understanding of forage products and forage management in Kansas. It is not meant to be a comprehensive text but rather a quick reference manual to aid seed retailers in making recommendations and forage producers in making appropriate decisions. This manual is meant to be easily read and understood; however, it is not 'light' reading. It is much too lengthy to read and absorb in one setting. The author recommends that the reader preview the preface of each chapter and then thumb through the various species discussed, reading those sections that hold particular interest in their operation or local geographical area.

It is very important to remember that no two operations are alike. What works for one producer may or may not work for his neighbor. In forage production there is no such thing as "one size fits all." Often, small differences in soil type, fertility and/or pH can result in significant differences in outcomes. In any agricultural endeavor, timeliness and attention to detail are key factors to a successful outcome.

Of course, producers can do everything correctly and, depending upon Mother Nature, still have a poor result. There are no guarantees in production agriculture. Therefore, if you are in a position to offer advice to producers, ask as many questions as necessary to determine their objective and their commitment to management. This will allow you to properly position products to meet their expectations. For example, should a producer want to introduce a legume to increase forage yield and quality in a hay meadow, red clover should be the first product considered. But, if soil pH is 6.3 or lower, Korean lespedeza or alsike clover will likely perform better. Whatever the circumstances, always be conservative in your recommendations. Remember, what you suggest will likely have an impact upon someone's livelihood.

Likewise, a producer searching for answers or solutions to maximize productivity and animal performance must do his or her homework. Do not rely upon one source of information or one agribusiness. Get to know your university extension specialists and county agents on a first-name basis. Also, the internet is a valuable resource that allows you to quickly search for information on any topic. Most seed companies understand that their success is dependent upon their customers' ability to compete and profit in production agriculture. Companies that understand this relationship are often an excellent source of information. And, publications such as *Forage for Kansas* provide information and the framework for successful forage production.

CHAPTER ONE: NATIVE GRASSES

There are many grass species native to Kansas. However, relatively few are of major economic importance. This section will be limited to those grasses which have a major impact on the Kansas livestock industry.

Many textbooks draw a line from north to south through the center of Kansas and refer to the area east of the line as the 'tall-grass region' and west of the line as the 'short-grass region'. In truth, the transition from tall-grass to short-grass is gradual, so much so, as to be almost imperceptible. For example, buffalograss can be found in the easternmost parts of the state and the tall-grasses appear even in the far western counties. While abrupt changes can and do occur due to differences in soils and precipitation, in central Kansas often the reason certain grasses dominate in a stand is more a function of management rather than a line drawn on a map.

Maximum production from any grass is dependent upon a working knowledge and understanding of the growth and development of that grass. Management must be geared to favor the key species in a grass stand in order to promote the persistence, vigor, and productivity of that grass. Especially on rangeland, this approach to grassland management is critically important.

Cattle are selective grazers when a choice is presented. They prefer some plants to others and will consume those plants first.

Because the grazing animal is selective, unmanaged grazing often results in stands dominated by grass species highly tolerant to grazing or by grasses with poor palatability. Those most preferred grasses are grazed continuously to the point where the carbohydrate reserves are depleted and they are unable to replenish themselves. Some of the most desirable species are the least tolerant to grazing. The height the growing points reach during the grazing season is highly correlated to grazing tolerance. Many grasses (such as blue grama) maintain their growing points near the soil surface until shortly before the onset of heading, while others (like switchgrass) push their growing points up within reach of grazing livestock relatively early in the growing season. If growing points are removed, any new leaf material must come from dormant buds at the base of the grass plants. This regrowth is powered by carbohydrate reserves in the roots. Should these reserves become depleted, obviously very little growth will occur.

There are a few tried and true management principles regarding grazing of native grasses. First and foremost is the principle of "controlled grazing." Controlled grazing refers to the practice of leaving enough of the current year's growth of the key species within a grass stand to maintain them from one year to the next in a healthy, vigorous state. Carbohydrate reserves stored in the roots of perennial grasses serve as the energy source to begin growth in the spring. If these reserves are depleted by over-grazing in the previous year(s), the grass will be low in vigor, less productive, more susceptible to other stresses, and in extreme cases, may die. However, if adequate leaf area is left at the end of the grazing season, carbohydrate reserves will be adequate to insure healthy, vigorous plants the following spring.

"Deferred grazing" is another management principle employed by the serious grassland manager. Its purpose is to allow the more desirable grasses to regain vigor and produce a seed crop. A full season deferment is highly recommended during the first growing season following brush-control measures, severe droughts, or overgrazing.

An "intensive management" system involves placing extremely heavy grazing pressure on the grass for relatively short periods of time and then rotating livestock to another pasture and allowing the first pasture to rest sufficiently to regain vigor. This practice favors maximum animal performance and utilization of available forage. However, it does require more input in the form of fencing, ponds or watering systems, and labor.

Intensive grazing is also often used to reduce competition from undesirable grasses within the stand or those invading the stand. For example, if tall fescue is invading a stand of native grass, intensive grazing of the infested area in fall, winter, and early spring with little or no pressure in late spring and the summer months, along with timely burns, can substantially slow and may even reverse the progression of the tall fescue.

Most of the warm-season native grasses are slowly established. The year of establishment usually results in very little top growth. Instead, all energies are devoted to establishing a root system. A common cause of stand failure is competition during the seedling year with weeds. Weeds compete with grass seedlings for space, sunlight, water and nutrients. It is very important to control weeds the year before attempting to establish native grasses. There are a few herbicides available for use on native plantings. However, at this time they are all broadleaf herbicides with little or no activity on the troublesome annual grasses.

When either purchasing or selling native grass seed, one should always deal on a pure-live seed (PLS) basis. Pure-live seed is derived from the following formula:

$$\frac{\%\ \text{Germination} \times \%\ \text{Pure Seed}}{100} = \%\ \text{PLS}$$

Example: Lot #001 Kaw Big Bluestem
60% Pure Seed, 80% Germination

$$\frac{60 \times 80}{100} = 48\%\ \text{PLS}$$

In other words in 100 pounds of this seed lot there are 48 pounds of actual viable seed. The warm season native grasses are very chaffy and extremely difficult to harvest and clean. Individual seed lots vary greatly in % PLS. By dealing on a PLS basis, both purchaser and seller are aware of the quality of a particular seed lot and both are able to make intelligent decisions on price and quantity needed to achieve desired stands.

LITTLE BLUESTEM

Little bluestem is a long-lived perennial warm-season native grass. A bunch grass with or without rhizomes, little bluestem grows to a height of two to four feet. Widely adapted, it can be found in every county in Kansas.

Excellent quality forage in early stages of development, little bluestem loses palatability and quality as maturity approaches. It is more drought tolerant than other grasses associated with the tall-grass complex i.e. big bluestem, indiangrass, switchgrass, etc., and persists longer under heavy grazing pressure on loamy to clay-textured soils in drier climates.

Establishment

Using a rangeland drill equipped with chaffy grass seed boxes, place seed at a depth of 1/4 inch on fine textured soils and 1/2 inch on coarse soils. A seeding rate of 4 to 6 PLS pounds per acre is usually adequate for normal plantings. If sowing highly eroded or disturbed land, increase seeding rate 50% to 100%.

The seedbed should be well prepared, weed-free and preferably pre-packed. In the western half of Kansas and on sandy sites in eastern Kansas, it is best to no-till seed into the standing stubble (16 to 18 inches) of sorghum or a non-volunteering crop. If seeding into a clean-tilled seedbed, it is often advantageous to cultipack or roll before and after seeding.

Unless seed production is the ultimate goal, pure stands are not generally recommended. Best forage production and animal performance is usually achieved when little bluestem is a 15% to 30% component of a mixture of grasses with which it associates in nature.

Stand establishment can be difficult. Best success is usually achieved with a March/April planting date; however, winter (dormant) seedings are usually successful and under irrigation a seeding date as late as June 1st may be attempted. Little bluestem is slow to establish, often taking two to three years to achieve desired stands. Weeds must be controlled during the establishment period either by mowing or using labeled herbicides. Once the stand is fully established, fields should be burned in early spring when new growth begins to appear. Burning aids in weed control, promotes tillering, and removes thatch.

Use and Management

Primary usage of little bluestem is for pasture, hay, and game bird cover. It is also an excellent conservation grass. Because of its growth habit and wide range of adaptability to climate and soil types, little bluestem has great merit for use in water channels, reclamation of mined lands and erosion control.

To maintain plant vigor, no more than 50% of the current year's growth should be grazed off. Grazing should be deferred or grazing pressure reduced substantially for 90 days every two to three years before seed maturity to allow the grass to regain vigor and produce seed.

If cut for hay, a 4 to 6 inch stubble should be left to promote recovery. The hay crop should be taken toward the end of July or the first of August to achieve best quality.

Varieties

Aldous is the preferred variety for central and eastern Kansas and also performs well in western Missouri, southeast Nebraska, and northeast Oklahoma. In western Kansas the variety Cimarron is preferred; however, in its absence, Aldous is a satisfactory substitute.

BIG BLUESTEM

Big bluestem is a dominant species of the tall-grass complex. A warm-season perennial native grass, big bluestem often grows to heights in excess of six feet. It is a bunch grass with a strong, deep root system with short rhizomes. Big bluestem is a highly praised forage grass, as well as an exceptional conservation species.

Best suited to loamy soils, big bluestem produces an abundance of wide bladed leaves highly palatable to all classes of livestock. In pure stands or mixtures, it provides the soil excellent protection against erosion. It builds soil organic matter rapidly through the decomposition of roots and tops. Big bluestem provides good quality forage throughout the summer months and makes fair winter roughage. For these reasons, big bluestem is often a preferred component species of grass mixtures when retiring cropland to meadows and pastures in eastern Kansas.

Establishment

Drill seed to a maximum depth of 1/4 to 1/2 inch on fine soils and 1/2 to 3/4 inch on coarse soils. A special rangeland grass drill with an agitator in the seed box and some mechanism to pull seed into the drop tubes is required. The drill should also be equipped with depth bands and press wheels. If planting into a clean-tilled seedbed, care should be taken in the preparation of the seedbed. Best success will be achieved if the seedbed is firm and seed is placed in a manner insuring good soil contact.

Generally, 5 to 8 PLS pounds per acre is adequate to achieve good stands. On highly eroded or critical sites, producers would be well advised to double the seeding rate.

In areas of Kansas with coarse soils, it is best to no-till seed into the standing stubble (16 to 18 inches) of sorghum or a non-volunteering crop, and on disturbed and/or highly sloping land, a mulch of wheat straw or prairie hay should be used. Best success is usually achieved with an early spring (March-April) planting date.

While big bluestem can be utilized very well in pure stands, it is most often seeded as a 20% to 50% component of mixtures of various other warm-season native grasses. Like most of the natives, big bluestem is slowly established, often taking two to three years to achieve desired stands. Weeds must be controlled during the establishment period either by mowing or using a labeled herbicide. Once big bluestem is fully established, a prescribed burning regime should be initiated. This will help control weeds and promote a vigorous stand.

Use and Management

Big bluegrass is a prize forage species; highly nutritious, it is widely used for pasture and hay. Big bluestem is one of the best quality forages in June, July, and August available to livestock producers in Kansas. It is also routinely utilized for wildlife habitat, reclamation of disturbed soils and erosion control.

Grazing should be deferred or grazing pressure substantially reduced every two to three years for 2 to 4 months before seeds ripen to allow plants to regain vigor and produce a seed crop. Responsive to nitrogen fertilization and irrigation, big bluestem is well suited to intensive grazing for short durations with long periods of rest between utilization.

Hay meadows should not be grazed during the growing season and only moderately during late fall and winter.

The best quality hay is cut in July; however, greater forage yield is achieved if cut in August.

Varieties

The variety Kaw is acceptable anywhere in Kansas. Kaw has superior leafiness and vigor when compared to common strains of big bluestem.

SAND BLUESTEM

Sand Bluestem is closely related to big bluestem. It is a vigorous perennial sod-forming grass with a deep fibrous root system and aggressive rhizomes. A single plant sometimes forms a dense colony as large as 30 feet in diameter. It is found on sandy soils throughout Kansas. It is the presence of the large, aggressive rhizomes that distinguish sand bluestem from big bluestem. The grasses cross-pollinate readily. Hybrids are intermediate to the parents and exhibit hybrid vigor. In fact, the improved variety Champ was selected from a hybrid population.

Sand bluestem begins growth in early-to-mid-spring and continues at a rapid rate throughout the summer. Plants generally obtain heights of 5 to 7 feet at maturity.

Establishment

Drill seed to a maximum depth of 1/2 to 3/4 inch using a rangeland grass drill. Plant 5 to 8 PLS pounds per acre into the standing stubble of sorghum or a non-volunteering crop. Most often, sand bluestem is seeded as a 20% to 30% component of a mixture of grasses with which it associates in nature.

Best success is usually achieved with an early spring planting date. Sand bluestem is slowly established, often taking 2 to 3 years for a desirable stand to develop. As with all the slowly established native grasses, competition from weeds is often the cause of stand failures. It is best to seed these grasses on fields that are relatively free of weed seeds; however, in Kansas this is seldom possible, and some form of weed control must be exercised the year of establishment. At present, the only method of annual grass control is mowing. Mowing is often ineffective in killing grassy weeds until these grasses put up seed stalks and by that time, most of the damage has been done.

Use and Management

Sand bluestem is utilized in the same manner as big bluestem. Sand bluestem performs well on sandy sites while big bluestem prefers loamy and clay textured soils.

After sand bluestem is fully established, fields should be burned at least every other year to control weeds and ensure vigorous stands.

Most often utilized as pasture, sand bluestem also makes good quality hay if cut before seed heads exert. For plant vigor and stand productivity, no more than 50% of the current year's growth should be grazed off. Grazing should be deferred or substantially reduced every three to four years for at least 120 days before seeds ripen. If cut for hay, a minimum stubble height of four inches should be employed.

Sand bluestem is an excellent conservation and wildlife grass. Its massive root system firmly holds the soil, preventing erosion, and its open sod provides exceptional game bird cover.

Varieties

Native stands are often the only source of seed. At present Woodward and Goldstrike are the two varieties most commonly recommended in Kansas.

SWITCHGRASS

Switchgrass is a vigorous perennial warm-season, sod-forming grass. It generally grows in association with the bluestems throughout the tall-grass region.

Switchgrass can be divided into two groups, bottomland and upland types. Bottomland types, such as Kanlow, occur in river bottom areas. They are taller, more robust and have a bunch type growth habit. Upland types are generally shorter and finer and have longer rhizomes that produce a spreading but open sod. Upland types, such as Blackwell, are usually 4 to 5 feet tall.

Switchgrass begins growth in mid to late spring and matures in mid to late summer. It is highly palatable to all classes of livestock until the onset of heading at which time quality rapidly declines.

Establishment

Drill seed to a maximum depth of 1/2 inch into a well-prepared firm seedbed or into the stubble of a non-volunteering crop. Two to four PLS pounds per acre are usually sufficient to achieve desired stands. If planting highly eroded or disturbed sites, increase seeding rates accordingly. Switchgrass is often recommended as a component of native grass mixtures for range plantings, but this is often a mistake. While switchgrass is not highly grazing tolerant, it will quickly dominate native stands in eastern Kansas under normal grazing management. Switchgrass grows very fast and reaches maturity quicker than the other tall grasses (big and little bluestem and indiangrass). It simply outgrows grazing animals' ability to consume it and as it approaches maturity, it becomes less palatable. Livestock leave it alone and consume the more palatable grasses in the stand.

This high quality forage is much more efficiently utilized in pure stands. Switchgrass responds well to nitrogen fertilizer and is well suited to an intensive management program tailored to its growth habit.

An early spring planting date is usually best; however, winter (dormant) seedings are also usually successful. One of the easiest natives to establish, it nevertheless requires two growing seasons before usable forage will be available. Weeds should be controlled in the establishment year, either by mowing or using labeled herbicides.

Use and Management

Switchgrass is used in both pure and mixed stands for grazing and hay, wildlife food and cover, as well as for grass-lined water channels and erosion control.

Switchgrass responds to proper grazing use and periodic deferments of 100 days at any point in the growing season. For maximum production of good quality hay, meadows should not be grazed in early spring and hay crops should be harvested at or before the onset of heading. Hay meadows can be grazed moderately in late fall.

Burning established stands will aid in weed control and prevent stands from becoming sod-bound.

Varieties

On upland sites in Kansas, Blackwell is the variety of choice. It is leafier and finer stemmed and more productive than native switchgrass.

On bottomland sites the variety Kanlow is recommended throughout Kansas.

INDIANGRASS

Indiangrass is a major range grass species in the eastern half of Kansas. It is widely distributed throughout the United States east of the Rocky Mountains. Indiangrass is a tall, vigorous bunch grass with wide leaves and a striking bronze or yellow plume-like seed head. Indiangrass forms rhizomes from which four-to-six-feet tall stalks arise. The species is highly regarded as excellent quality forage utilized for pasture or hay. Indiangrass, of all the tall warm-season native grasses, has the greatest potential to be utilized east of the Great Plains states.

Indiangrass thrives on deep, moist soil varying from heavy clays to coarse sands. It tolerates moderate soil acidity and moderate salinity. While tolerant to flooding, it is perhaps the least drought tolerant of the warm-season native grasses.

Establishment

Drill seed with a rangeland drill equipped with chaffy grass seed boxes into a pre-packed, clean-tilled seedbed or the stubble of a non-volunteering crop. Usually, a planting rate 5 to 6 PLS pounds per acre is adequate to achieve desired stands. If seeding highly eroded land or disturbed sites, increase seeding rates 50% to 100%. In Kansas, best success is usually achieved with an April or early May planting date.

Indiangrass can be, and often is, planted in pure stands. However, it is more often seeded as a 20% to 50% component of mixtures of warm-season natives, such as big and little bluestem, switchgrass, sideoats grama, etc.

Seedlings show relatively good vigor when compared to other natives, making indiangrass one of the easiest warm-season natives to establish. However, it will be the second growing season before usable forage is available. Weeds should be controlled during the establishment year, either by mowing or using labeled herbicides.

Use and Management

Indiangrass is excellent quality forage used either for grazing or haying. It is also routinely utilized for erosion control, game bird cover, and grass-lined water channels.

Indiangrass can tolerate close grazing much better than the other grasses of the tall-grass complex, and responds well to fertilization under irrigated or high rainfall conditions. For these reasons, indiangrass is often planted and managed in pure stands. In an intensive grazing study conducted in Bourbon County, Kansas, steers grazing pure indiangrass averaged daily gains of nearly 3 pounds from mid-June through mid-September. (Bourbon Co. SCS 1984).

While indiangrass is grazing tolerant, it nevertheless is one of the first species to disappear if grasslands are mismanaged. Because indiangrass remains palatable for the entire growing season, livestock tend to search it out and consume it in preference to other grasses in the association. On sites that have a high percentage of indiangrass in the plant community, it is the key management species. On less productive sites, its presence indicates that the range is in good to excellent condition.

As with the other warm-season native grasses, burning in early spring is a good management practice.

Varieties

Osage is the recommended variety for eastern Kansas, western Missouri, and northeast Oklahoma. In central Kansas, Osage and Cheyenne are both acceptable. In more western areas, Cheyenne is preferred.

SIDEOATS GRAMA

Sideoats grama is a warm-season native perennial grass. It possesses an extensive fibrous root system. This grass is very winter hardy and drought resistant and is adapted to a wide range of soil and climatic conditions. Most commonly found in mixture with blue grama, buffalograss, and little bluestem, it predominates in this association on shallow soils, steeply sloping lands, deep sand and exposed sites.

Sideoats grama is prized for wildlife plantings and is an exceptional conservation grass. Seedling vigor is excellent compared to other warm-season native grasses and establishment is relatively easy.

Growth begins in mid-spring and maturity is reached in late July or early August. Plant height at maturity is approximately three feet. Sideoats grama produces leafy forage that is palatable to all classes of livestock. Good quality hay may be produced if cut in a timely manner.

Establishment

Using a rangeland drill equipped with chaffy grass seed boxes place seed to a maximum depth of 1/2-inch in heavy soils and 3/4 inch in sandier soils. Drill at a rate of 5 to 6 PLS pounds per acre. If seeding into clean-tilled seedbed, care should be taken in its preparation. The seedbed must be firm, weed free, fine on top and loose enough below the surface to allow easy penetration of seedling roots. It must not be powdery. If the ground must be worked more than what would be considered ideal, it is best to wait for a rain to settle and firm the surface.

With an April planting date, seeds will normally germinate within 21 days of planting. One of the easiest of the native to establish, a forage crop can sometimes be achieved the year of establishment. Weeds must be controlled during the first growing season, either by mowing or by using labeled herbicides.

Use and Management

Native stands are most often utilized for pasture and less commonly for hay. Because of its wide adaptability, rapid establishment and deep extensive root system, sideoats grama is an excellent conservation grass. It is routinely used on eroding and disturbed sites and grass-lined water channels. Sideoats grama is also prized for wildlife plantings. Seeds are relished by songbirds and small mammals use it for cover and food.

Sideoats grama responds well to rotational deferred grazing systems. These management systems allow the most efficient and effective utilization of available forage. For stand maintenance and productivity, sideoats grama should be allowed to mature and make a seed crop every two to three years.

Prescribed burning of sideoats grama is very beneficial in increasing productivity.

Varieties

The variety El Reno may be successfully utilized throughout Kansas. It has demonstrated superior forage and seed yields when compared to local strains.

BLUE GRAMA

Blue grama is a warm-season native perennial grass found throughout the short-grass region from Texas to Canada and is also important in parts of the desert grasslands. Blue grama is usually found in association with buffalograss and in the northern area of its adaptation with the wheatgrasses. In the more southern areas, blue grama can and often does occur in near pure stands.

A typical short-grass, blue grama seldom grows taller than 12 to 20 inches. In semi-arid areas, blue grama is often utilized as a turf grass. It forms a dense sod and its narrow leaves can make an attractive lawn. For turf, blue grama is usually seeded in mixture with buffalograss. A mixture of one part buffalograss to two parts of blue grama produces an extremely drought tolerant and low maintenance turf.

Adapted to a wide range of soils and climate, blue grama produces a highly nutritious forage during the summer months. Forage remaining in the fall makes good winter pasture if allowed to cure standing. Tolerant to alkaline soils, blue grama is the most drought tolerant of the major grasses on the Great Plains.

Establishment

Using a rangeland drill equipped with chaffy grass seed boxes, place seed no deeper than ½ inch at a rate of 2 to 4 PLS pounds per acre. Most often blue grama is seeded as a 10% to 30% component of a mixture of grasses with which it associates in nature. Drill into a well-prepared, weed-free, firm seedbed or into the stubble of a non-volunteering crop. Seed can and often does germinate rapidly under ideal field conditions. However, seedling vigor is poor and establishment can be difficult. Allow 2 to 3 years to establish a desirable stand. In Kansas, April and May seeding dates are generally recommended. Winter (dormant) plantings have also been successful. Weeds must be controlled during the establishment period, either by mowing or using a labeled herbicide.

Use and Management

Adapted to a wide range of soil types, blue grama is used in range seedings, conservation plantings, wildlife habitat plantings, roadsides, and as a turf grass in semi-arid regions of the Great Plains.

Native stands are most often utilized for range and pasture. It produces an excellent quality forage and with the exception of buffalograss, it's the most grazing tolerant grass on the Great Plains.

Blue grama is routinely used for revegetating abandoned croplands and disturbed areas. Birds relish the seeds and small mammals consume seed heads and plants.

Since growing points are at or near the soil surface for most of the season, blue grama tolerates close grazing. For best performance, defer grazing every two to three years during the growing season and graze no more than 50% of the current year's growth.

Varieties

Native harvest is often the only source of seed. The variety Lovington is most often recommended in Kansas. When compared to native seed stock, Lovington demonstrates superior seedling vigor.

BUFFALOGRASS

Buffalograss is a deeply-rooted native perennial grass, important in the short-grass region from Texas to South Dakota. It is found primarily on soils with fairly high clay content and does not perform well on sandy soils. This grass spreads rapidly by means of surface runners and forms a dense, matted growth 5 to 8 inches tall. Buffalograss is predominantly dioecious (individual plants are either male or female), with male and female plants occurring in nearly equal frequency.

Growth begins in mid-spring and continues all summer. The forage is highly palatable to all classes of livestock. Buffalograss is fairly easy to establish and spreads vigorously under grazing pressure. It withstands prolonged heavy grazing better than any other grass native to the Great Plains states. In fact, in western Kansas on range heavily grazed every year, it commonly survives as a nearly pure stand.

Establishment

Drill 6 to 8 PLS pounds per acre at a depth of 1/2-inch. Generally, an April or early May seeding date is best in Kansas. Buffalograss seed is encased in hard burs, with one or several viable caryopsis (grains) to a bur. Seed dormancy can be a serious problem in stand establishment. However, soaking seed in a 0.5% solution of $KN0_3$ for 24 hours, followed by chilling at 40^{0} F for six weeks will effectively break dormancy. Use of such "primed" seed will greatly enhance the success of establishment.

In range seedings, buffalograss is usually seeded as a 20% to 50% component of a mixture of grasses with which it associates in nature.

Use and Management

Buffalograss is widely used in the short-grass region for pasture and erosion control. It is also utilized for turf in semi-arid regions.

Buffalograss withstands heavy grazing and is ideally suited to intensive grazing systems. It has an excellent reputation as cured winter feed. Livestock usually harvest less than 50% of the current year's growth because this grass grows so close to the ground. However, continuous close grazing can result in weakened plants and reduce the next season's production.

Varieties

The varieties Texoka and Sharps Improved are recommended throughout Kansas and can be readily substituted for one another. Both varieties produce a higher percentage of female plants in their offspring than do native stands, thus ensuring better seed and forage production. Forage yields of both varieties exceed those of native stands.

Bison is a newer variety from Oklahoma that has demonstrated impressive performance in Kansas.

EASTERN GAMAGRASS

Eastern gamagrass is a warm-season perennial bunch grass native to the eastern United States. Eastern gamagrass produces the majority of its growth from mid-April through mid-September. It begins growing earlier than other warm-season natives, and is at its peak when cool-season grasses (like tall fescue) are dormant or growing very slowly.

Eastern gamagrass is characterized by numerous short, well-developed rhizomes. Most of the leaves originate near the base of the plant. Leaves are wide (1 to 1 1/2 inches) and long, forming an upright and arching canopy. Individual grass clumps can reach a diameter of 4 feet with seed heads growing on stems 3 to 9 feet tall.

Eastern gamagrass is a relative of field corn and is much more responsive to fertilizer than other warm-season native grasses. It requires a minimum pH of 5.5 and performs best when soil pH is 6.0 to 7.0. The plants will utilize large amounts of water, but due to its remarkable root system, will survive and even thrive on droughty sites and during dry growing seasons.

Establishment

The major problem associated with Eastern gamagrass production is the difficulty of establishment. There are only about 7,500 seeds in a pound of gamagrass seed. The large seed size makes it necessary to plant more pounds per acre than is required for the other warm-season native grasses. Planting is usually accomplished with a corn planter in 30-inch rows. Seeds should be placed at a depth of 1/2 to 1 1/2 inch.

Recommended seeding rates range from 8 to 10 PLS pounds per acre. Some producers will plant twice, splitting the original 30-inch rows. This results in roughly 15-inch rows and may result in better distribution of seed across the field and quicker ground cover, which aids in weed control. Whatever you do – do not plant in a checkerboard fashion. Individual plants can be 4 feet in diameter with huge crowns. Bouncing over these crowns will likely loosen your teeth and damage equipment.

The other major challenge in establishment is seed dormancy. Eastern gamagrass seed requires stratification to break dormancy. The two most commonly used techniques to break gamagrass dormancy are (1) to plant stratified seed (wet-chilling of seed at 35° F for 10 weeks) when soil temperatures reach 65° F in April or May, or (2) to plant unstratified seed in the winter (November thru February) and rely on natural soil temperatures to break seed dormancy.

Use and Management

Eastern gamagrass is highly palatable and nutritious forage that, properly managed, will produce total seasonal yields in excess of 6.5 tons of dry matter per acre. It may be grazed, harvested for hay or put up as silage. As with other grasses, the quality of the forage declines as plants mature.

While eastern gamagrass is sometimes included in mixtures with other native grasses for rangeland or pasture plantings, it is best to sow it in pure stands. Eastern gamagrass maintains excellent palatability throughout the grazing season. Cattle are selective grazers and they will often ignore other grasses in the association and graze the gamagrass to extinction. Pure stands of eastern gamagrass allow producers to design production systems tailored to the growth and development of this grass that meet their needs for high quality forage.

Initial forage harvest should be delayed until at least the year following establishment and may need to be deferred an additional year to ensure an adequate stand. Forage removal as hay or silage should occur prior to boot stage, when plants are 24 to 36 inches tall. Initial grazing should not occur until plants are 18 to 24 inches in height. An 8-inch stubble should be maintained to ensure that adequate leaf material is present to support regrowth. Harvests of regrowth should occur at four-to-six-week intervals following the initial harvest. If grazed, a rotational stocking system should be employed. This allows the forage to be grazed to the proper stubble height and provides the rest periods necessary for optimum regrowth and stand maintenance.

To determine the need for nitrogen, phosphorous, potassium and lime, producers should submit soil samples for testing. A rule of thumb is to request recommendations for corn silage and apply these materials at the recommended rate to Eastern gamagrass. To reduce competition from weeds, no nitrogen should be applied at seeding. After seedlings emerge, an application of 30 to 40 pounds of nitrogen per acre will promote seedling growth.

On established stands, nitrogen fertilizer should be applied at a rate of 50 pounds per acre as new growth begins in the spring. An additional application of 50 pounds nitrogen should be made following each subsequent forage harvest, to promote regrowth and enhance forage quality. After the final harvest in early September, no nitrogen should be applied. However, phosphorous and potassium should be top dressed in accordance with soil test recommendations.

As new growth appears in the spring, established stands of eastern gamagrass should be burned. Burning removes any thatch buildup, helps in weed control, and promotes new tillers.

Varieties

Pete is a composite of seed collections made in Kansas and is generally the preferred variety in Kansas and Missouri. PMK-24 is very similar to Pete and is an acceptable substitute. Iuka is a variety with more southern germplasm and is well suited to Oklahoma, Texas, and Arkansas.

WHEATGRASSES

Western wheatgrass is a cool-season native perennial sod-forming grass. It has a deep fibrous root system with vigorous rhizomes that effectively protect the soil from wind and water erosion. It is a major rangeland species in the northern and central Great Plains. Western wheatgrass is widely adapted to varying soil and climatic conditions. Where soil is favorable and moisture plentiful, western wheatgrass may be found in nearly pure stands. On upland sites it is commonly found in association with the grama grasses.

Western wheatgrass starts growth early in the spring and produces a high protein forage relished by all classes of livestock. Plants at maturity attain a height of 2 to 3 feet. This grass goes dormant during the hot/dry months, but makes further growth in early fall. Established stands of western wheatgrass can endure long periods of severe drought, tolerate flooding, soil salinity, poor drainage, and withstand heavy grazing pressure.

Establishment

Drill seed to a maximum depth of 1/2 inch on fine-textured soils and 3/4 inch on sandy soils. A seeding rate of 5 to 8 PLS pounds per acre is usually adequate to achieve good stands. The seedbed should be well-prepared and firm. Seed may be planted in late winter/early spring or late summer. In Kansas, a late August/early September planting date is preferred because it allows a longer establishment period before the onset of hot/dry summer weather. Also, late summer plantings have less competition from weeds. Seeds usually take 21 to 56 days to germinate and seedlings are not vigorous. Western wheatgrass can take 2 to 3 years to fully establish. Because of this slow establishment, it is not generally recommended to plant western wheatgrass in pure stands. It is most often seeded as a 15% to 25% component of mixture of other adapted grasses.

Use and Management

Widely used for pasture and hay, western wheatgrass is also an excellent grass for erosion control, critical area stabilization, grass-lined water channels, and surface mine reclamation.

Well suited to intensive grazing systems, western wheatgrass provides high quality forage when the warm-season natives are at their worst. For maximum production on rangeland, defer grazing every few years at least 90 days before seed heads exert. Protect this grass from grazing throughout the growing season if it is managed for seed harvest or hay. Graze only moderately during the dormant period.

Varieties

The variety Barton is generally recommended throughout Kansas. Seed and forage yields of Barton have proven superior to native stands.

There are a few other species of wheatgrass that are not native to North America but are sometimes recommended in rangeland and conservation plantings in Kansas.

INTERMEDIATE & PUBESCENT WHEATGRASS

Intermediate wheatgrass is a tall-growing cool-season grass with moderately vigorous creeping rhizomes. It generally establishes quickly and is very productive. It flowers about three weeks later than smooth brome and provides an extended period of summer use. It generally stays green following mild frosts in the fall. Intermediate wheatgrass does not tolerate flooding or saline soil conditions. Its area of adaptation is from Nebraska to Manitoba, Canada and west to Washington and California. In Kansas, its use should be restricted to the northern-most counties. Generally, I do not recommend intermediate wheatgrass anywhere in Kansas. Our hot and dry summers severely limit the persistence of intermediate wheatgrass and we have better alternatives.

Pubescent wheatgrass is a subspecies of intermediate wheatgrass that is usually distinguished from intermediate by pubescence on spikes, seed, and occasionally on leaves. Pubescent wheatgrass has demonstrated better adaptation than intermediate wheatgrass to droughty, infertile soils and saline areas. While it is a better fit in Kansas than intermediate wheatgrass, stand persistence is still an issue.

TALL WHEATGRASS

Tall wheatgrass originated from southern Europe and Asia Minor. It is the latest maturing wheatgrass adapted to temperate areas in western North America and one of the most productive. This species is commonly used on range sites receiving at least 20 inches of annual precipitation or supplemental irrigation. Tall wheatgrass produces good forage yields in areas that are too alkaline or saline for most other grasses. Although it becomes coarse as it approaches maturity, it remains green three to six weeks longer than most other cool-season grasses. This makes it a valuable source of pasture or hay during the summer, when forage is often in short supply.

Tall wheatgrass has good seedling vigor and is relatively easy to establish. When utilized for pasture, it is recommended that a 6 to 7 inch stubble of be left at season's end to prevent animals from grazing too close the following spring. Grazing should not be initiated until 6 to 8-inches of new growth have accumulated above the stubble.

Jose is the variety most commonly utilized in Kansas.

CHAPTER TWO: INTRODUCED GRASSES

As a group, the introduced grasses are some of the most important forage and turf grasses in Kansas. These grasses introduced from around the world, through passage of time and breeding efforts, have become naturalized to those sections of the country to which they are adapted. Kansas, due to its geography, is often referred to as a transition state. Many grasses have a limited adaptation to Kansas. However, relatively few are well adapted over a wide area of the state. The introduced grasses include both warm and cool-season grasses. One of the most important differences between warm and cool-season grasses is the way in which they conduct photosynthesis. Photosynthesis is the process by which plants convert atmospheric carbon dioxide into carbohydrates and oxygen. This process can be credited for all life on earth. The atmosphere contains very little carbon dioxide (0.03%); the success of a plant is dependent upon its ability to collect and use carbon dioxide. The warm-season grasses have what is referred to as a C4 photosynthetic process, while the cool-season grasses adapted to Kansas use a C3 photosynthesis system.

C4 (warm-season) plants are more efficient in gathering carbon dioxide than are the C3 (cool-season) plants when both are growing at optimum temperature. The optimum temperature for growth of cool-season grasses is 65° to 75° F while it is 90° to 95° F for warm-season grasses. Warm-season grasses are also more efficient in water use than the cool-season grasses. These differences explain why warm-season grasses are more productive in the hot/dry summer months while cool-season grasses are most productive during the cool and moist spring and fall months.

In Kansas, the cool-season introduced grasses are utilized to a much greater extent than are the warm-season introduced grasses. Perhaps the best suited of the cool-season grasses are smooth bromegrass and tall fescue. Other cool-season grasses with more limited adaptation include orchardgrass, reed canarygrass, timothy, Kentucky bluegrass, and the ryegrasses.

As a group these cool-season introduced grasses are highly responsive to fertilization and proper management. It is important to understand the growth and quality characteristics of these grasses in order to design the proper forage program for a particular ranch operation. For example, if fall pasture is needed or if forage needs to be stockpiled for winter pastures, tall fescue is a better selection than either smooth bromegrass or orchardgrass. On the other hand, if spring pasture or quality hay production is the producer's goal, both smooth bromegrass and orchardgrass are superior to tall fescue.

Seeding legumes in combination with the cool-season grasses is an excellent management practice. Legumes can increase carrying capacity, improve quality and digestibility, as well as lower fertilizer costs. However, if the two species are to coexist, the grass/legume mixture must be managed properly. Proper management would include liming to near neutral pH, annual application of potassium and phosphate fertilizer, heavy grazing pressure or clipping of grass in early spring to allow the legume the luxury of less competition and either rotational intensive grazing or periods of grazing deferment when necessary.

Examples of warm-season introduced grasses planted in Kansas include the old world bluestems, weeping lovegrass, crabgrass, and bermudagrass. None of these grasses are utilized to a great extent at present; however, as bermudagrass cultivars with improved winter hardiness become available, this species could become a valuable forage in eastern Kansas and under irrigation in more western sections of the state.

SMOOTH BROMEGRASS

Smooth bromegrass, native to Europe and Asia, is adapted to most temperate climates. Deeply rooted, smooth bromegrass survives periods of drought and extremes in temperature. During dry summer periods, it becomes dormant until the return of cooler temperatures and fall moisture. Best suited to deep fertile soils of well-drained silt loam, smooth bromegrass can nevertheless be utilized on a wide range of soil types. It is a leafy, medium-tall growing, sod-forming perennial grass. It spreads underground by rhizomes and is readily propagated by seed.

Forage quality of smooth bromegrass is usually superior to other cool-season grasses (tall fescue, orchardgrass, timothy, etc.). A disadvantage of bromegrass is its slow recovery after cutting, which contributes toward slight regrowth and poor seasonal distribution of yield. Regrowth is slow because growing points are removed by cutting and consequently regrowth must come from below ground nodes. In order to keep smooth bromegrass productive year after year, it must be skillfully managed, particularly in the fall of the year. Close grazing of smooth brome through the fall may result in thin, low vigor stands the following spring. Properly managed, this superior grass will remain productive for several generations of Kansas farmers.

Establishment

A moist, fertile, firm seedbed is required to establish smooth bromegrass. It can be successfully seeded in spring or late summer/early fall. In Kansas, late summer plantings are preferred. This allows smooth brome to become fully established the following spring before the onset of another hot summer season. Another advantage of late summer plantings is that there is usually less weed pressure in autumn than spring.

A seeding rate of 15 to 20 pounds per acre of good quality seed is usually sufficient to achieve desired stands. Drill seed to a depth of 1/2 to 3/4 inch into a well-prepared, clean-tilled seedbed. Small grains are sometimes used as a companion crop with smooth bromegrass. However, if companion crops are to be successfully utilized, seeding rates must be adjusted so that they do not become overly competitive.

Use and Management

Smooth bromegrass is highly regarded for use as irrigated and non-irrigated pasture, quality hay production, soil conservation and critical area stabilization, and in grass-lined waterway seedings.

New stands of smooth bromegrass should be protected from grazing until the grass is well established. Broadleaf weeds can be controlled with low rates of 2, 4-D after grass seedlings have at least 5 or 6 leaves. The annual grasses can be suppressed by mowing. Mow at a height of 4 to 6 inches and do not remove more than 3 inches of smooth bromegrass leaves.

Dense stands become sod-bound in three to five years without a perennial legume in the stand or nitrogen fertilizer. Maintenance of established stands is usually dependent upon yearly application of fertilizer and adequate moisture. An annual application of 90 to 100 pounds actual nitrogen in winter with phosphate and potassium included as recommended by a soil test is sufficient for hay or seed production.

Bromegrass is well suited to intensive grazing systems in the spring of the year. Hay meadows should not be grazed prior to cutting and only moderately in the fall. Good quality and high forage yield will be achieved when smooth bromegrass is cut in early bloom stage.

Varieties

Kansas is considered one of the main smooth brome seed production areas in the United States. This seed is considered "common" or "variety not stated." This common seed stock produced in Kansas performs very satisfactorily and is generally preferred over seed of Canadian origin.

TALL FESCUE

Tall fescue is a cool-season perennial grass introduced from Europe. Generally, tall fescue is well adapted to the humid temperate areas of the United States and the world. While tall fescue grows and performs best on good, moist soils that are heavy to medium in texture, it exists and performs relatively well on soils that vary from strongly acid (pH 4.7) to alkaline (pH 9.5). It thrives and conserves soils on thin, droughty slopes, and forms dense sods on poorly drained soils where few other cool-season grasses survive. The massive root structure of tall fescue is generally credited for its adaptation to many varied soil types. The roots of tall fescue decrease soil density, improve soil structure, and firmly hold the soil, thus reducing erosion.

Cattle grazing tall fescue occasionally will exhibit physiological disorders, i.e. fescue toxicity (fescue foot), poor animal performance (summer syndrome), and fat necrosis. Extreme symptoms of fescue foot and fat necrosis are easily distinguished, but it is not known if they are different responses caused by the same agent(s), or an extension of the less acute response of summer syndrome.

Summer syndrome is associated with the symptoms of rough hair growth, reduced rate of gain and/or milk production, rapid breathing, increased body temperature, and a general unthrifty condition during the warmest grazing season (July and August). Summer syndrome is seldom fatal; however, due to the large number of cattle grazing tall fescue, summer syndrome has a much greater economic impact than fescue foot or fat necrosis.

The occurrence of summer syndrome corresponds in time (the hot summer months) to the increased accumulation of certain alkaloids and an endophytic (inside plant) fungus, E. Typhina. A positive correlation has been found between the endophtic fungus and alkaloids (organic bases) linking them to the occurrence of summer syndrome. Thus, the endophyte and/or alkaloids appear to be casual agent(s).

Establishment

Tall fescue plantings normally are made during fall from mid-August through mid-October or during the spring from March through May. Seeding rates for forage purposes vary from 15 to 25 pounds per acre. Seed should be planted at depths of 1/4 to 3/4 inch by either broadcast or drill equipment into a clean, weed-free, firm seedbed. Often small grains are utilized as companion crops when establishing tall fescue. This generally works very well as a grain crop may be harvested the first summer and tall fescue forage the following season. However, normal seeding rates of small grains should be cut in half to insure that the tall fescue will compete.

Planting legumes in mixture with tall fescue is highly recommended. However, it can be difficult to maintain legumes in a lush, vigorous stand of tall fescue. It is usually necessary to place heavy grazing pressure on the grass in early spring or to clip the grass back to allow the legume the luxury of less competition. Often liming and fertilizing with phosphate and potassium at rates necessary to maintain the companion legume will be sufficient for the tall fescue. However, applications of nitrogen, as circumstances warrant, will be necessary to keep the grass green, succulent, and highly productive.

Use and Management

Tall fescue is widely used for pasture, hay, turf, and conservation purposes. Tall fescue occupies an estimated 30 to 35 million acres in pure and mixed stands within the United States, making it the predominant cool-season grass species.

Factors that govern the quality of tall fescue forage include the age of the leaves, fertility of the soil, and the season of the year. Quality is improved if the plants are kept grazed closely, clipped to prevent accumulation of old leaves, and properly fertilized or grown in association with a legume. Quality is lowest during summer, intermediate in spring and highest during the fall season. The development of cultivars with a low level of alkaloids and the elimination of seed transmission of the endophytic fungus should permit realization of tall fescue's potential for meeting the nutritional needs of the livestock industry.

Tall fescue is grazing tolerant and best animal performance is achieved if grazing pressure is sufficiently heavy enough to prevent the accumulation of older leaves.

If cut for hay, best quality is achieved before seed heads exert.

Varieties

Most tall fescue in Kansas and Missouri is Kentucky 31, which can be credited with saving millions of tons of soil from erosion and restoring productivity to a vast region of over-farmed and mismanaged crop and rangeland. Ky 31 will continue to be used by both turf and livestock industries. However, with the discovery of the link between the endophytic fungus and summer syndrome, much demand has been generated for varieties free of the fungus. Examples of these varieties include Kentucky 32 and Atlas.

The most exciting innovation in tall fescue breeding has been the introduction of endophyte-enhanced varieties. These varieties have the fungus, but it has been genetically modified and does not adversely affect animal performance. In fact, the endophyte makes the fescue more stress tolerant and disease resistant, which prolongs stand life. Pennington Seed Company released the first variety,

MaxQ, and has enjoyed the advantage of exclusivity and uniqueness of the product in the marketplace. However, recently several other companies have introduced endophyte-enhanced varieties. Estancia tall fescue looks very promising. When compared to MaxQ, it is advertised as being softer (more palatable) and more productive.

ORCHARDGRASS

Orchardgrass, a native of Europe, has been grown in North America for over 200 years. It has spread through a large area of the United States, where it has become a very important forage species. A cool-season perennial grass, orchardgrass is utilized for hay and pasture. While orchardgrass is more tolerant to heat than timothy or Kentucky bluegrass, it is less so than tall fescue or smooth bromegrass. It is moderately shade tolerant and routinely found growing where there is reduced light, such as orchards.

Soil requirements of orchardgrass are less exacting than those of either timothy or smooth bromegrass. Orchardgrass will persist on shallow, rather infertile soil and be modestly productive. However, it responds well to fertilizer, especially nitrogen, and becomes very competitive when nutrients are available. At high rates of nitrogen fertilization, orchardgrass is among the most productive of the cool-season grasses. Hay yields in excess of 4 tons per acre can be expected when orchardgrass is properly fertilized and growing conditions are favorable.

A high quality forage, orchardgrass, when in the vegetative growth stage, approaches the feeding value of alfalfa. At full bloom, orchardgrass possesses approximately half the nutritional value of alfalfa.

Establishment

Orchardgrass is relatively easy to establish. A seeding rate of 12 to 15 pounds per acre is usually sufficient to achieve desired stands. Seed should be drilled into a well-prepared, firm seedbed at a depth of 1/2 inch. Oats are frequently used as a companion crop. The oats are harvested for hay, silage, or grain during the summer and the orchardgrass is harvested the following year.

It is a good practice to seed legumes in combination with orchardgrass. When seeding this combination, it is best to plant a full rate of legume seed and cut by 1/3 to 1/2 the orchardgrass seeding rate.

Use and Management

Orchardgrass is leafy, productive, and adapted to a wide range of environmental conditions. Once established it will survive many years if properly managed. It is well suited for pasture, hay, greenchop and silage, and it can be utilized alone or in combination with legumes.

Growth characteristics of orchardgrass make it well adapted for early spring pasture and better suited to rotational grazing rather than continuous grazing. When grazed continuously, plants become severely weakened by the frequent removal of leaf tissue.

Ladino clover is excellent for use in combination with orchardgrass for pasture. The clover will provide nitrogen for the grass, and if properly managed, both species will coexist and remain productive for a number of years.

Varieties

In eastern Kansas and Missouri the varieties Hallmark and Potomac are most widely planted. These varieties are considered early in maturity and possess good winter hardiness.

RYEGRASSES

The ryegrasses are adapted to temperate regions around the world. The two most important forage species are annual ryegrass (L. multiforum Lom.) and perennial ryegrass (L. perenne L.). The ryegrasses are all more or less interfertile (able to cross-breed) which results in great variation in plant types. Tetraploids (double the number of chromosomes) of annual and perennial ryegrasses have been developed in breeding programs and are widely utilized in forage and turf applications.

Perennial ryegrass use in the United States is limited by its lack of winter-hardiness and relatively shallow root system, which makes it less persistent than other temperate grasses, such as tall fescue, bromegrass, or orchardgrass.

Annual ryegrass, also known as Italian ryegrass, is used as a winter annual and is highly productive.

The tetraploids are generally intermediate between annual and perennial ryegrass in their persistence and forage yield. These cultivars generally have fewer but larger tillers and wider leaves than their perennial parent.

All ryegrasses perform best on fertile, well-drained soils. However, they can grow and thrive on soils so wet at certain times of the year that few other grasses will survive. Both perennial and annual ryegrasses are considered bunch-type grasses and provide excellent quality forage.

Establishment

The ryegrasses are fairly easy and quick to establish. It is best to begin with a clean, well-prepared seedbed. However, they may be established in standing stubble or existing grass sod without difficulty. If sowing into a perennial grass sod, the sod should be mowed or grazed short or desiccated with an herbicide prior to seeding.

A seeding rate of 15 to 25 pounds per acre of good quality seed is usually sufficient to achieve good stands. Drill seed to a depth of 1/2 to 3/4 inch. If sown with a legume, ryegrass seeding rate is generally cut to 8 to 12 pounds per acre. Broadcast seedings are routine and usually successful. However, root development and soil penetration are less rapid and seedling growth is frequently slowed.

In Kansas, late summer or early fall seeding dates are most often recommended. This timeframe allows the ryegrass to become fully established and utilized during the late fall, winter and spring seasons.

Use and Management

Both annual and perennial ryegrasses are considered excellent quality forage almost everywhere they are grown. Only the lack of adaptability and persistence limits their use in pastures. Additional uses are for hay, silage, soil conservation, and turf. Because of generally high digestibility, they are suitable for all classes of ruminants (including lactating dairy cows).

In Kansas, annual ryegrass is sown in late summer or early fall and generally grazed all winter and into the spring. In the spring it is either grazed-out or taken as a hay crop. As temperatures warm in late spring or early summer, the annual ryegrasses make a seed crop and die.

The perennial ryegrasses may be sown in late summer or early fall or early spring. Perennial ryegrass is generally not as quick to establish, nor is it as productive in the year of establishment. It does, however, generally persist for 3 to 6 years (unless a severe drought or winter occurs).

Varieties

The vast majority of annual ryegrass planted in Kansas is "common" or Gulf annual ryegrass. DH-3 is an improved annual with better winter hardiness and forage yield. Improved varieties of annual ryegrass are often utilized in cover crop systems in Kansas.

The most popular variety of perennial ryegrass is Linn.

TIMOTHY

Timothy is a bunchgrass with a shallow, fibrous root system. It is best adapted to the cool and humid areas of the northeastern United States. Its usage in Kansas is limited to the eastern most portion of the state and has steadily declined over the years to the point of only minor importance. While smooth brome, tall fescue, and orchardgrass have largely replaced timothy in Kansas, it still maintains popularity, particularly for horse hay and pasture.

Timothy lacks the drought and heat tolerance necessary to make it a long-lived perennial in Kansas. It thrives on heavy, moisture-retentive soils; however, it does not tolerate water-logged soils as well as reed canarygrass, red top, or tall fescue. Timothy is a relatively noncompetitive grass, and as such, is excellent for use in combination with legumes. It increases yield and adds ground cover to legume mixtures without depressing legume yield or persistence.

Timothy is considered a very palatable grass. Its forage quality usually ranks higher than tall fescue, reed canarygrass, or orchardgrass, and slightly lower than smooth bromegrass. In a grass/legume hay mixture, timothy's late maturity is considered an advantage, as it is relatively young and tender at harvest.

Establishment

In Kansas, timothy should always be planted in combination with legumes. When fall seeded, 2 to 4 pounds per acre of good quality seed is all that is required. If seeded in the spring, 4 to 6 pounds per acre of timothy in the mixture is recommended. Timothy has approximately 1,200,000 seeds per pound, so heavier seeding rates are not necessary due to the small seed size. Because of the small seed size, planting depth is critical – do not drill too deeply! Prepare the seedbed as you would when planting alfalfa and drill the mixture no deeper than 1/4 to 1/2 inch.

Use and Management

Timothy is grown primarily for hay and is also routinely utilized in mixtures for pasture. Timothy is widely renowned for its high quality and palatability, particularly in the horse industry. Good quality and maximum forage yield is achieved if cut at or before the onset of heading.

Timothy is a very desirable pasture species. A mixture of 4 pounds of timothy and 1 pound ladino clover is an excellent horse pasture mixture for eastern Kansas and Missouri. It is best not to seed timothy in mixture with aggressive and/or grazing tolerant grasses such as tall fescue. However, mixed stands of timothy, red top and/or orchardgrass in combination with alfalfa, birdsfoot trefoil or ladino clover can persist and remain productive for a number of years.

Varieties

The variety Climax is recognized as being more persistent in Kansas than common.

Clair is a variety that should be considered when available. It is earlier maturing than the northern varieties and produces larger aftermath yields.

REED CANARYGRASS

Reed canarygrass is a tall, wide-leafed, sod-forming, perennial cool-season grass. It spreads underground by short, scaly rhizomes and forms a dense sod in well-managed pure stands. Individual plants exhibit considerable difference in agronomic characteristics, suggesting that there is much potential for selection of improved cultivars.

Reed canarygrass is one of the highest yielding perennial grasses available for grazing or haying. However, the lack of palatability or of selection by livestock when a choice is presented is why this extraordinary grass has not become a leading forage in its area of adaptation.

Especially well adapted to wet, marshy land, Reed canarygrass is very tolerant of flooding. It will also grow on upland soils and shows considerable resistance to drought. It produces lush growth from early spring through summer, but produces little growth in fall and winter. Reed canarygrass develops a sod sufficiently firm to support livestock even on wet land. Likewise, farm machinery is supported on sod in areas that would be impassable before the establishment of Reed canarygrass.

In Kansas, Reed canarygrass is best adapted to the northern third of the state. However, it has been successfully utilized throughout the state.

Establishment

Reed canarygrass seed is often low in germination. Sow 5 to 10 pounds per acre of good quality seed (75% germination or better). Either spring (March/April) or late summer (August/September) planting dates are satisfactory if sufficient moisture is available. Plants must be well established before the onset of winter because Reed canarygrass seedlings are more susceptible to winter-kill than are those of most other cool-season grasses. Seed should be drilled or broadcast and cultipacked into a well-prepared seedbed. Take care not to incorporate seed too deeply; ideally seed should be covered with 1/4 inch of soil. Seed and seedlings can withstand 35 or more days of flooding.

Use and Management

Reed canarygrass is used for pasture, hay, silage, and soil stabilization. For the best quality pasture, allow Reed canarygrass to reach a height of 8 inches before stocking with livestock and place heavy enough pressure on the grass to keep it shorter than 30 inches. Rotational grazing with heavy pressure for short periods provides the best utilization of pasture.

Best quality hay will result when the first crop is taken before heading. With adequate nitrogen and moisture, three hay crops per season may be obtained.

Reed canarygrass makes good quality and palatable silage when harvested in a timely manner. Preservation of Reed canarygrass silage is excellent, provided the crop is harvested by the early heading stage and is finely chopped.

Varieties

Very few varieties of Reed canarygrass have been developed and as a result most seed planted is of "common" or "variety not stated" origin. When available, the low alkaloid varieties Palaton and Venture are recommended in Kansas.

KENTUCKY BLUEGRASS

Kentucky bluegrass, despite its name, is not native to the United States. It was most likely introduced to this country in the mid 1700's. At one time it was the predominant cool-season pasture grass utilized in the United States. However, its usage has declined over the years to the point where it is of major importance only in the north central and north eastern states. In Kansas, usage of Kentucky bluegrass should be limited to the north eastern most counties. It lacks the drought and heat tolerance necessary to survive as a long-lived perennial in other areas of the state.

Kentucky bluegrass performs best on well-drained, productive soil with pH values higher than 6.0. It is a short to mid-height, cool-season perennial grass. It spreads readily by rhizomes and forms a dense sod. Rhizomes develop buds during late summer and fall and give rise to new tillers throughout the cool fall and spring seasons. Kentucky bluegrass is most productive in the fall of the year, moderately productive in the spring, and poor in the summer.

A prized turf grass, Kentucky bluegrass, properly managed, produces a beautiful fine-textured lawn.

Establishment

In Kansas, Kentucky bluegrass should be sown in late summer or early autumn. Usually 8 to 10 pounds of good quality seed per acre is adequate to achieve good stands. When seeding bluegrass in combination with a legume, cut the seeding rate to 4 1/2 to 7 pounds per acre. Best success in stand establishment is achieved by planting seed with a Brillion drill or similar equipment into a well-prepared firm seedbed. However, seed is often broadcast and then harrowed or cultipacked to insure good soil contact. Care must be taken not to incorporate seed too deeply.

Use and Management

Generally speaking, there are several cool-season grasses better adapted and more productive in Kansas than Kentucky bluegrass. Even in northeastern Kansas, where Kentucky bluegrass performs satisfactory, it is not as productive as smooth bromegrass, tall fescue, orchardgrass, or for that matter, the warm-season natives.

If Kentucky bluegrass is to be utilized in Kansas, it should be used in combination with legumes, such as ladino clover or birdsfoot trefoil. Kentucky bluegrass and legume mixtures provide good quality pasture. Highly grazing tolerant in its area of adaptation, Kentucky bluegrass often survives in near pure stands with a minimum of management. In fact, pastures must be skillfully managed to prevent Kentucky bluegrass from quickly dominating a grass stand. Because of its ability to recover from buds on rhizomes (even if tops are sheared off by sharp hooves) Kentucky bluegrass makes a good horse or sheep pasture.

Kentucky bluegrass should not be used if haying is the producer's goal. Many other grass species will provide superior hay yields with equal or better forage quality.

Varieties

When sowing Kentucky bluegrass for pasture, the common types or older varieties, such as Park, are better than the newer turf-type varieties. The "common" varieties and strains are taller growing and generally require less nitrogen fertilizer.

BERMUDAGRASS

Bermudagrass probably originated in southeast Africa. In Kansas, bermudagrass has been successfully utilized in about a twenty-county area of southeast and east-central Kansas. The western boundary is determined by inadequate moisture and the northern boundary by temperature and winter hardiness. The release of the variety Guymon by Oklahoma State University and the USDA ARS has greatly altered the parameters of bermudagrass production in Kansas. Guymon is winter-hardy even in the northernmost counties of Kansas. With the advent of varieties such as Guymon, bermudagrass could become a major warm-season forage in eastern Kansas.

Bermudagrass is a warm-season perennial, sod-forming grass. It excels when temperatures approach 100° F. Bermudagrass will grow on any moderately well-drained soil provided it has an adequate supply of moisture and plant nutrients. Although it tolerates flooding for long periods of time, it makes little or no growth on water-logged soils.

Bermudagrass is as excellent quality forage if it is kept young and growing. The first cutting of hay may have a protein content as high as 19%. Subsequent cuttings of fertilized bermudagrass (cut every 25 to 30 days) usually averages 12% crude protein.

Establishment

Until recently, sprigging bermudagrass was the only method to successfully establish this grass in Kansas. Guymon was the first commercially available variety that can be established with seed and have sufficient hardiness to survive the first winter.

When sprigging bermudagrass, 10 to 15 bushels of sprigs per acre is required and a rate of 20 bushels per acre is more desirable. The seedbed should be well-prepared, firm and moist, but not wet. Drying of sprigs before or after planting will cause decay and molding, which prevents growth. Use only freshly dug sprigs or those that have been properly stored. If sprigs are to be transported or stored for any period of time, shade and moisture are necessary to prevent drying and deterioration. Commercial sprigging machines are the best method of establishing bermudagrass. The sprigging machine should be set to plant the sprigs about 2 inches deep. The soil should be firmed around the sprigs at planting with press wheels and/or a heavy roller.

If planting seed, 6 to 8 pounds per acre is sufficient to achieve desirable stands. A Brillion drill or comparable equipment is best. Seed should be placed no deeper than 1/2 inch and best results are usually achieved when the seedbed is kept moist during establishment, but not water-logged.

Whether using sprigs or seed, an early-to-mid-May planting date is advised in Kansas. Soil temperatures must be high enough to promote rapid growth of seedlings or sprigs. However, it is very important to get plants established before the soil becomes hot and dry.

Use and Management

Bermudagrass is very desirable forage used either for haying or grazing. Utilized in pure stands or in combination with legumes, bermudagrass provides good quality forage during the warm spring and summer months. When managed for pasture, it is important to remember that forage quality deteriorates rapidly with maturity. Continuous grazing results in spot or patchy grazing. Rotational, intensive grazing is preferable. Ideally, bermudagrass should be heavily grazed for ten days, and then deferred for three weeks. If the grass gets ahead of grazing livestock, cut it for hay. Do not let bermudagrass mature.

When managing bermudagrass for hay production, the important considerations are proper fertilization and timely cutting. Permitting the grass to grow longer than 6 weeks between cuts lowers forage quality dramatically. Hay conditioners will hasten curing so hay can usually be baled 24 hours after each cutting. Nitrogen fertilizer (50 to 100 pounds per acre) should be applied after each cutting through August to maintain forage yield and protein content. Phosphate and potassium should be applied annually with the first application of nitrogen as recommended by a soil test.

Varieties

Midland 99 and Hardie are the varieties most commonly recommended for Kansas. Both possess good cold tolerance and are highly productive. However, they are slowly established and must be sprigged at considerable expense to the producer.

Wrangler bermudagrass has unquestionably the best cold tolerance and winter hardiness of all the seeded cultivars currently available. In Oklahoma, Wrangler does not produce the forage yields that other improved varieties provide. However, as it is taken north in Oklahoma, the production gap narrows substantially. In Kansas, Wrangler survives in areas where other cultivars will not, and it will yield considerably more forage than common bermudagrass.

CRABGRASS

Crabgrass is a warm-season annual grass that grows in every state in the contiguous United States. It grows best when temperatures range from 80° to 100° F. Production is best at a day length of twelve hours or more, the longer the day the greater the production. Best performance is usually achieved on lighter soil types. It does not perform well on very tight clay loam or heavy clays.

While crabgrass is moisture-loving, it does not tolerate soils that remain waterlogged for extended periods of time. It grows well with 24 to 60 inches of annual rainfall if other conditions are favorable. Production falls off dramatically when annual rainfall does not exceed 20 inches. Maximum production requires adequate moisture; however, crabgrass is very drought tolerant. It will survive periods of extreme drought, but production will suffer.

Crabgrass, in its area of adaptation, is highly productive. It will produce forage yields of 5,000 to 10,000 pounds per acre per year under favorable conditions. Properly managed, crabgrass produces excellent quality forage. Typically, gains of 2 to 2.5 pounds per day have been realized from well-managed crabgrass pasture.

Establishment

Crabgrass can be established easily from seed planted from mid-winter through early summer. Crabgrass is typically either sown in a firm, fine seedbed or over-seeded in wheat or rye. When planting in a clean-tilled field, placing the seed on a rolled (cultipacked) seedbed and then rolling the surface once again to insure good soil/seed contact is best. When over-seeding into a winter annual (such as wheat) seed should be sown from mid-to-late winter so that freezing and thawing and snow will help pull the seed down into good soil contact.

Seeding rate recommendations range from 2 to 6 pounds per acre. Generally, good success is usually achieved with a rate of 3 to 31/2 pounds per acre on clean seedbeds and 4 to 6 pounds per acre broadcast into winter cereals. Seed should be placed on the surface of the soil or no deeper than 1/2 inch. Once the stand is established, it may be maintained for several years. Crabgrass is indeterminate and can produce new tillers and ripen seed simultaneously from June to fall. This allows crabgrass to build a seed bank, which results in the next season's volunteer stand.

Use and Management

Crabgrass makes a highly palatable pasture of high quality forage. Production and daily gains are highest in a rotational grazing program that is well managed. Crabgrass should be allowed to accumulate to a height of 10 to 16 inches and then grazed back to a 3 to 6 inch height. Recovery periods of 4 to 6 weeks (depending upon the season and rainfall) are adequate.

Crabgrass also makes excellent quality hay. It should be cut when it is 12 to 24 inches tall leaving a 3-to-6-inch stubble and a green leaf on most stems. This stubble height is extremely important as new growth is initiated from the green leaf. Well managed crabgrass meadows will produce forage yield and quality superior to those of bermudagrass.

In order to maintain productive stands of crabgrass, it is necessary to till the soil in years subsequent to establishment. Tillage to a depth of 2 to 4 inches is adequate and may be accomplished with equipment such as discs, field cultivators, and sweep plows. Spring tillage prior to crabgrass emergence is best, as it promotes a more rapid and uniform stand of grass.

Crabgrass responds to good soil fertility, especially nitrogen. To promote production and forage quality, nitrogen fertilizer should be applied at a rate of 100 pounds per acre. If splitting nitrogen applications, research demonstrates a linear response to nitrogen application in excess of 200 pounds per acre. Crabgrass produces well on a wide range of soil pH. However, best production is achieved when soil pH is in the range of 6.0 to 7.2.

Varieties

Red River is a cultivar developed by the Noble Foundation of Ardmore, Oklahoma. Red River is extremely productive and more aggressive than local strains. Quick N Big is a new variety that looks promising for Kansas producers as well.

CHAPTER THREE: WARM-SEASON ANNUALS

The warm-season annual grasses utilized for pasture, greenchop, silage, or hay are an excellent supplement for perennial cool and warm-season meadows and pastures. Most cool-season grasses become semi-dormant during the hot summer months, and most warm-season perennials provide forage relatively low in protein at this time of the year. The warm-season annuals are most productive during the months of July, August, and September and provide a high quality forage.

Warm soil temperatures (70° to 80° F) are essential for rapid, uniform emergence, growth, and development. Once established, the summer annuals adapt to both hot and dry conditions.

There are several members of the sorghum and millet families utilized as warm-season annuals. The sorghums utilized include grain sorghums, sorgos, grass sorghums, and sudangrass. The most widely used millets are pearl millet and in more northern areas, foxtail millet. The importance and prominence of forage sorghums increased rapidly in the 1960's with the advent of improved hybrids. Likewise, improvement of pearl millet cultivars and hybrids has led to its increased usage while most other millets are declining.

Teff grass is a recently introduced summer annual forage that is gaining in popularity. It is high yielding forage with exceptional quality.

The forage sorghums and foxtail millet are single-cut products, but the other summer annuals may be harvested multiple times with a potential forage yield of 2 1/2 tons per acre or more per harvest. Plants are difficult to cure because of their thick stems; even sudangrass and pearl millet with thinner stems may be difficult to cure because of the large amount of forage to be dried. Sunshine, wind, and the use of a crimper help reduce the time required to dry these products.

Livestock are the most economical harvesters of any forage. However, grazing summer annuals demands considerable management skill. Planting rates and stocking rates must be carefully adjusted to efficiently utilize forage and maximize animal performance. The threat of prussic acid poisoning in the sorghum family and nitrate poisoning in both sorghums and millets are real concerns. Grazing often leads to waste due to trampling or fouling by excreta. All of these issues must be addressed by the producer/manager in order to effectively utilize warm-season annual grasses for pasture.

Utilizing sorghums and pearl millet as silage or greenchop for dairy cows is an increasingly common practice. Management of forage sorghums for silage is similar to that of corn silage. Proper moisture content (65% to 70%), length of cut, packing, and avoidance of air and moisture are critical in the making of good quality silage. Sorghums become progressively less digestible as they mature. Thus, it's imperative to harvest when whole plant moisture is in the range of 65% to 70%. Silage inoculants may be necessary as the relatively high moisture of these forages is favorable for butyric acid producing bacteria. These bacteria result in bad smelling, unpalatable feed. Silage inoculants will decrease butyric acid production and speed-up fermentation time.

TEFF GRASS

Teff is a self-pollinated warm season annual grass that is thought to have originated in Ethiopia. It performs well in environments ranging from drought stress to water-logged soil conditions on diverse soil types. It is highly susceptible to frost at all stages of growth and will not survive at temperatures below freezing. While teff is considered drought tolerant, it is most productive in regions having at least 20 inches of annual rainfall or supplemental irrigation.

Teff grass is excellent quality forage, comparable to timothy in nutritive value. It is also high yielding forage capable of producing forage yields of 1 1/2 to 2 1/2 tons of dry hay within 45 to 55 days after planting. Typically, in eastern Kansas, teff's total seasonal forage yield will range from 6 to 7 tons per acre.

Establishment

Stand establishment is perhaps the most critical component in the successful utilization of teff, which is very small seeded, averaging about 1.3 million seeds per pound (compare to alfalfa which has 220,000 seeds per pound). For this reason, coated seed is much preferred over raw seed, as coated seed allows growers to utilize their own planting equipment and achieve a uniform distribution of seed across the field.

The recommended seeding rate is 8 to 10 pounds per acre. A Brillion seeder is the optimum way to plant teff grass. However, broadcast seeding is usually successful provided the surface is rolled or packed after sowing. Conventional or no-till drills may be utilized if the drill is equipped with a small seed box attachment. The surface of the ground being sown must be very firm and the seed should not be placed deeper than 1/4 inch. Achieving a good, uniform stand is critical to maximize production. If working the ground results in a powdery surface, it is best to wait for a rain to firm the surface prior to planting.

As a warm-season annual, teff does not grow well in cool soils. Soil temperatures should be at least 65° F and warming prior to planting (last half of May). Teff germinates rapidly under warm conditions (within 4 or 5 days if moisture is adequate). However, most of the plant growth during the first two weeks is devoted to establishing its root system and not to top growth. Therefore, early competition from weeds can be a serious issue in stand establishment. A broadleaf herbicide may used, but it should be delayed until the teff plants have 5 to 7 leaves. Following the initial root growth period, teff is very aggressive in its growth and will out-compete most weeds.

Use and Management

Teff is best suited for haying operations. However, cattle, horses, or sheep can graze it. When pasturing, teff is most productive in rotational grazing systems with periods of rest between grazing to allow the plants to recover and accumulate growth. Avoid grazing teff until the root system has fully developed.

Harvesting teff as hay will result in maximum forage production and return per acre. When haying, leave a minimum of 4 inches of stubble to promote quick regrowth. Teff is fine stemmed, leafy, and very palatable to all classes of livestock. Single-cut yields are often 2 to 2 1/2 tons per acre with forage quality comparable to timothy. In Kansas, producers may expect to harvest teff hay three times in a 'normal' growing season, with a total production of 6 to 7 tons per acre. Teff is considered a low input crop, requiring minimal fertilization. Total seasonal nitrogen needs range from 70 to 90 pounds. While nitrate accumulation issues have not been reported in teff, it is wise to err on the side of safety. Split applications of nitrogen (35 to 45 pounds) following each cut throughout the growing season will keep the protein level good in subsequent cuttings and reduce the risk of nitrate accumulation.

Teff does not have any issue with the prussic acid poisoning that is common in forages in the sorghum family.

Varieties

Teff is a relatively new forage crop in Kansas and exhibits great potential in the state. Currently, the most respected varieties include Velvet Brand and Tiffany Brand teff grass.

SUDANGRASS

There are two classes of sudangrass: varietal sudangrass (Piper and Greenleaf) and hybrid sudangrass (Trudan).

Varietal sudangrass is fine-stemmed and leafy. It is primarily utilized for grazing and hay. Varietal sudangrass should be drilled at a rate of 10 to 15 pounds per acre at a depth of 1 inch. It can be planted anytime after soil temperatures exceed 65° F (usually late May or early June in Kansas). As long as soil moisture is adequate, the sudangrasses may be planted throughout the summer months, but expect it to take four to six weeks before usable forage is available.

While varietal sudangrass produces less forage yield than other members of the sorghum family, they are generally of superior quality. Varietal sudangrass ranks lower in danger of prussic acid accumulation than other sorghums; however, the danger does exist and care must be taken when managing as pasture.

Use a soil test to determine needs for potash and phosphate fertilizer and apply 40 to 60 pounds actual nitrogen preplant and 40 pounds after each cutting or grazing.

If utilized for grazing, allow sudangrass to reach a height of 18 to 20 inches and stock heavy enough to keep all plants from obtaining a height of 3 feet. Rotational grazing systems that allow sudangrass rest periods to recover and regenerate growth are the most productive and efficient utilization of this forage. Prussic acid can be concentrated in new or young growth. Therefore, delaying the introduction of livestock until plants have obtained sufficient growth is a key component in grazing management.

If cut for hay, highest quality will be achieved by cutting before plants begin to head. Cut each time growth reaches 40 inches for maximum production and highest quality. Always leave a minimum of 6 inches of stubble to promote rapid regrowth. North-south rows may yield as much as 5% more than east-west rows when harvesting frequently at an immature stage of growth.

Management and usage of hybrid sudangrass is very similar to that of varietal sudangrass. However, hybrids have the potential to yield more forage per acre. There is a danger of prussic acid accumulation in this crop and caution must be exercised, particularly in periods immediately following drought or light frosts. Prussic acid is concentrated in new growth. If the crop grows continuously, there is little danger; however, should growth stop and then begin again...look out.

Hybrid sudangrass should be drilled at a rate of 15 to 25 pounds per acre into a moist, firm seedbed. Soil temperatures should exceed 65^{0} F before attempting to establish this grass. Manage for hay or pasture as you would varietal sudangrass. Hybrid sudangrass offers an excellent compromise, as it provides nearly as good forage quality as varietal sudangrass and greater yield. Generally, hybrid sudangrass yields less than hybrid sorghum/sudangrass, but is superior in forage quality.

HYBRID SORGHUM/SUDANGRASS

The hybrid sorghum/sudangrasses are versatile, high-yielding forages. These products are best utilized by haying, silage, ensilage, or by greenchopping. They produce greater yields of dry matter per acre than other sudan-type crops. However, the forage produced is generally of lesser quality. Daily gains of livestock grazing hybrid sorghum/sudangrass have been disappointing when compared to the other summer annuals. This is due in large measure to the thick stems, which translate into high crude fiber and low palatability. Large stem diameters can be tempered somewhat by high plant populations. This crop should be drilled into a well-prepared, firm seedbed at a rate of 25 to 40 pounds per acre. Soil temperature should exceed 65° F before attempting establishment. Ideally, seed should not be placed deeper than 1 1/2 inch.

When managing hybrid sorghum/sudangrass for pasture or hay production, fertilize with 60 to 75 pounds actual nitrogen preplant. Then apply 30 to 50 pounds nitrogen after each cutting or grazing. If managing for silage 75 to 100 pounds nitrogen should be applied preplant. Potash and phosphate materials should be supplied as determined by soil tests.

Prussic acid accumulation is a definite threat to grazing livestock. Likewise, nitrate poisoning is possible when high rates of nitrogen fertilizer are applied pre-plant.

The introduction of sorghum/sudangrass hybrids with the brown mid-rib trait (BMR) is perhaps the most significant innovation in recent forage breeding. BMR hybrids have greatly reduced lignin content in the plant. Lignin is the woody component of the plant cells that is generally regarded as the primary factor limiting forage fiber digestion. BMR sorghum/sudangrass has averaged roughly 19% increased feed value when compared to traditional sorghum/sudangrass. The BMR trait results in much more palatable forage that is more digestible and results in better utilization of available forage (less waste).

With sustainable production systems gaining in popularity, hybrid sorghum/sudangrass has gained an additional role as a cover crop. It is unrivaled for adding organic matter to worn-out soils. As a fast growing, heat-loving summer annual, hybrid sorghum/sudangrass can smother weeds, break the cycle of many diseases, suppress nematode populations, and when mowed once, can penetrate compacted subsoil.

Generally, hybrid sorghum/sudangrass grows to a height of 6 to 10 feet with long leaves and stalks up to 1/2 inch in diameter and possesses an aggressive root system. These features combine to produce a large biomass (usually 2 to 3 tons per acre).

When compared to non-hayed fields, haying when stalks reach 3 to 4 feet in height increases root mass dramatically. Cutting as hay results in much deeper root penetration and can result in fracturing compacted subsoil layers.

Hybrid sorghum/sudangrass seedlings, shoots, leaves, and roots secrete allelopathic compounds that suppress many weeds such as velvetleaf, crabgrass, barnyardgrass, green foxtail, smooth pigweed, common ragweed, redroot pigweed and purslane. Additionally, the high seeding rates and fast, vigorous growth of hybrid sorghum/sudangrass can effectively smother most weeds.

Planting hybrid sorghum/sudangrass instead of a host crop is an effective way to disrupt the life cycle of many diseases, nematodes, and other pests. Studies have demonstrated that this crop produces natural nematicidal compounds. Timing of harvest and tillage is critical in the suppression of nematodes. The hybrid sorghum/sudangrass must be tilled (incorporated in the soil) while still vegetative. Otherwise the nematicidal effect is lost. The best strategy seems to be to hay the crop at its appropriate stage of maturity and then, after 40 to 45 days of regrowth, cut it again, but this time, incorporate it into the soil with tillage. This strategy will take full advantage of hybrid sorghum/sudangrass' disease and nematode suppression capabilities, while adding significant organic matter to the soil. It should also allow enough time to breakdown the residue (before this activity is limited by cold temperatures) and provide a good seedbed the following spring.

FORAGE SORGHUMS

In eastern Kansas, the forage sorghums are used almost exclusively for the production of silage. However, in central and western Kansas, forage sorghums are more often utilized as high quality baled feed.

Prepare a firm, moist seedbed and plant 4 to 6 pounds seed per acre for dryland production and 6 to 10 pounds under irrigation or in high rainfall areas. Usually planted in rows, forage sorghum should be planted at a depth of 1 to 1 1/2 inches in fine-textured soils and 1 1/2 to 2 inches in sandy soils. Apply 70 to 100 pounds actual nitrogen and potassium and phosphate as recommended through soil tests.

When haying, best forage quality is achieved if harvest occurs in the boot to early-head stages. While forage yield may be greater at later stages of maturity, the quality of the forage will be dramatically reduced.

Because forage sorghums have rather large stems, the temptation to increase seeding rates to moderate stem diameters is always a consideration. However, this is often a serious mistake. Planting forage sorghums at high plant populations makes this crop much more susceptible to temperature and moisture stress. When the plants begin to "fire" growth slows or ceases which may result in dangerous levels of nitrate accumulation. A much better management practice is to plant brown mid-rib (BMR) hybrids. The BMR hybrids produce a much softer (more palatable) and more digestible forage regardless of stem diameter.

When used for silage production, these sorghums generally compete very well with corn silage from the standpoint of yield and quality, and they do so with considerably less water. For silage, forage sorghums should be harvested when seed are in the milk-to-dough stage, before leaves reach senescence. Earlier or later harvest will affect quality, energy, and yield. Length of silage cut at the soft dough stage should be 1/2 inch, and at medium to hard dough, the cut should be 3/8 inch.

Milk production is greater with hybrids having good grain to roughage ratios. The highest quality silage is obtained when grain makes up 25% to 35% of the dry matter yield.

There are a few open-pollinated varieties still grown in Kansas. Examples include Hegari, and the cane types, such as Rox Orange and Early Sumac. While these products serve a purpose, the hybrids are preferred due to higher productivity.

CAUTION: Never feed to or allow horses to graze any sorghum crop!

Members of the sorghum family have been reported to cause "cystitis" syndrome in horses. Cystitis (inflamed urinary bladder) is characterized by uncontrolled urination, abortion or stillbirths in mares, and by the inability to coordinate rear quarters. Cystitis affects only horses and is of no significance in ruminants. Horses are also susceptible to the prussic acid and nitrate-poisoning problems more commonly associated with the sorghum family. For these reasons, it is best to avoid sorghum forages in the feeding of horses.

HYBRID PEARL MILLET

Hybrid pearl millet is very fine-stemmed when compared to the sorghums. This leafy forage is well suited to grazing or haying. While hybrid pearl millet generally yields less forage per acre than sorghum crops, it nevertheless has produced as much or more beef per acre than the sorghums in university trials. What hybrid pearl millet lacks is quantity, it makes up for in quality.

Producers like the fact that there is little or no danger of prussic acid accumulation in hybrid pearl millet.

Fertilize with potash and phosphorous according to a soil test and apply 45 to 70 pounds actual nitrogen preplant. Following each cutting or period of grazing, apply 30 to 40 pounds additional nitrogen.

Prepare a firm, moist seedbed and sow 15 to 20 pounds per acre of good quality seed. Depth of seed should not exceed 1 inch on fine soils and 1 1/2 inches on sandy soils. Hybrid pearl millet will not tolerate cold wet soil conditions. It therefore, should not be planted until soil temperature approaches 70° F (usually after June 1st). Hybrid pearl millet is susceptible to the triazine herbicides where the sorghums are not. Thus, previous cropping and herbicide history should be considered prior to planting.

When haying hybrid pearl millet, leave 8 to 10 inches of stubble to promote a rapid, leafy regrowth. Cut whenever forage obtains a height of 3 1/2 to 4 feet for maximum quality. Unlike the sorghums, hybrid pearl millet relies upon above ground buds on the stem to initiate regrowth. Mowing too closely will severely limit pearl millet's ability to recover.

Hybrid pearl millet utilized for pasture should be allowed to reach a height of 12 to 18 inches before stocking with livestock. The most efficient utilization of available forage occurs when pearl millet is grazed intensively and then allowed to rest sufficiently for regrowth to occur.

FOXTAIL MILLET

In Kansas, foxtail millet is grown as forage that is capable of producing good quality hay or pasture in a short amount of time and with minimal inputs. Unlike the sorghums, foxtail millet does not accumulate prussic acid and is tolerant to high pH soils. It is well suited to less fertile soils and poorer growing conditions, such as very hot temperatures and low rainfall. Because of its rapid growth and shorter growing season, foxtail millet is ideal for emergency or double-cropping situations.

Foxtail millets form slender, erect leafy stems varying in height from 1 to 4 feet. Small convex seeds are encased in hulls and are borne on a spike-like compressed panicle (seedhead).

Foxtail millet requires warm weather and grows quickly in the hot summer months. While it has a low water requirement, it does not recover well from drought conditions because it has a shallow root system. In Kansas, foxtail millet generally requires 60 to 70 days to reach its peak forage yield in haying operations. It makes very good quality hay for cattle or sheep.

Establishment

Foxtail millet is very sensitive to cold temperatures and planting should be delayed until the soil warms to a minimum of 65° F at a depth of 1 inch. It can be planted as late as mid-August in Kansas, as long as there is sufficient moisture for establishment. Conventional grain drills equipped with a small seed box (alfalfa box) are best due to the small seed size. Drill 15 to 20 pounds per acre at a depth of 1/2 inch. Planting too deeply will result in poor emergence.

When planting into recently harvested wheat fields, the wheat straw should be mowed or baled or light tillage should be used to reduce the stubble. Millet seedlings need sunshine and standing wheat stubble will shade the millet plants. This may adversely affect early growth and stand establishment.

Use and Management

Foxtail millet can be productive on less fertile soils, but it does respond to nitrogen and phosphorus fertilizers. Generally, 45 to 75 pounds nitrogen and 30 to 60 pounds phosphorus per acre are sufficient to produce a good quality hay crop. Producers should windrow foxtail millet at the boot-to-early-head stage. If cut later, nutritive value will be lower, but forage yield may be higher. Hay yields normally range from 1 1/2 to 2 1/2 tons per acre. Regrowth after cutting is slight to nil, so it is a single-cut crop. Millet hay can be dangerous to horses if fed in large quantities due to a glucoside (setarian) that can damage kidneys, bones, and joints. Therefore, foxtail millet forage should only be fed to cattle and sheep.

Foxtail millet makes very good pasture for cattle and sheep, but grazing should be delayed until plants have achieved 12 to 16 inches in height. Millet is shallow rooted and is fairly easily pulled from the ground by grazing livestock until it is sufficiently anchored by a well-developed root system.

While there is no danger of prussic acid, nitrate poisoning can be an issue when either grazing or haying this crop.

Varieties

In Kansas, the vast majority of foxtail millet planted is the variety German Strain R developed by Texas A&M many years ago. German Strain R is an early maturing variety with consistent, proven performance.

Golden is another variety sometimes planted in Kansas. When compared to German Strain R, Golden is earlier in maturity and generally produces lower forage yield. Golden also has waxier leaves. This waxy characteristic may allow the forage to retain its quality longer in situations where harvest or baling is delayed.

CORN AND SORGHUM SILAGES

Silage has long been utilized as an integral component of beef and dairy cattle rations. Often the decision to ensile a crop is made late in the season as the result of drought or early frost. This versatility allows great flexibility in cropping practices and allows the producer/manager to successfully and profitably salvage a crop. Harvest of corn or sorghum as silage results in the maximum yield of nutrients and provides a good quality, highly palatable feed. Once a silage crop is harvested, it may be stored for long periods of time, provided it is ensiled at the proper moisture (65% to 70%) and protected from spoilage losses. The addition of silage to feed rations reduces the problems of "off-feed" and founder.

While corn and sorghum silages are similar, there are some different management aspects and qualities. Generally, corn silage provides better animal performance than sorghum silage. This is due in large measure to the higher energy value of corn silage. Forage sorghum silage will have 80% to 90% the energy value of corn silage per unit of dry matter, while estimates for sorghum/sudan hybrids range from 60% to 80% the value of corn silage. However, when soil and climatic conditions cause moisture stress (as is the case in most areas of Kansas in the summer), dry matter yields favor the sorghum silages. Even though the sorghum silages contain less digestible energy than corn silage, yields of digestible energy will most often favor the sorghums in Kansas.

Maximum dry matter yield of corn silage will be obtained at the dent-to-glaze stages of kernel maturity. Digestibility and animal intake will also be near maximum at these stages. Forage sorghums should be harvested as silage promptly when whole plant moisture is in the range of 65% to 70%. Hybrids with "dry stalks" should be cut in the soft dough stage of kernel maturity to obtain the optimum dry matter yield and protein content. Those hybrids with higher stalk moisture content will have to be delayed for harvest until the grain head has colored to achieve the 65% to 70% moisture level.

The basic management practices such as length of cut, packing, avoiding contaminates such as dirt and dust, and covering to prevent exposure to air and moisture apply to both corn and sorghum silage. Corn and sorghum silages that contain considerable grain are often referred to as "high-energy." These high-energy silages may be used as the only source of forage in cattle and sheep rations. However, both are relatively low in protein content. Several research studies have demonstrated a gain advantage when supplementing rations containing these forages with protein.

One of the most important considerations in silage management is the preservation of quality. Forage losses occur during harvest, storage, and feeding. In an attempt to reduce storage-related losses, many additives have been developed. However, no matter how good an additive may be, there is no substitute for good management. Ensiling at the proper moisture greatly reduces storage losses. Forage that is too dry results in the trapping of oxygen, which causes molding of silage and excessive spoilage. If ensiled at high moisture levels, forage losses due to seepage and excessive fermentation occur. Other practices that reduce spoilage include fine chopping to aid in packing, rapid filling of silos to lessen exposure of surfaces to oxygen, use of a distributor to reduce the amount of segregation of light and heavy particles as silage is blown into the silo, proper packing of silage in the silo, and covering the surface with plastic to reduce surface spoilage.

PRUSSIC ACID POISONING

Prussic acid (HCN) or hydrogen cyanide has long been recognized as a potentially devastating problem in the pasturing of livestock on sorghum forage. Symptoms of prussic acid poisoning include nervousness, increased rate of respiration, trembling, blue coloration of mucous membranes, spasms, convulsions, and death. Prussic acid is concentrated in new growth of leaves and shoots. As the plants mature, prussic acid content declines due to the relative increase in stem, leaf, sheath, and midrib. The regrowth after cutting, intensive grazing, or after a mild frost can be very high in prussic acid because it is essentially all new leaves and shoots. Likewise, the growth that occurs immediately after a period of drought may have dangerously high levels of prussic acid. Generally, as long as growth is continuous, the danger of prussic acid poisoning is relatively slight; however, should growth be interrupted and then begin again, extreme caution must be exercised.

The risk of prussic acid poisoning can be reduced by sound management practices. Do not begin grazing of these products until 18 to 20 inches of forage accumulation has occurred. If the crop is stunted by drought or frost, do not graze or chop until the crop has recovered (several days after the first flush of growth). Do not pasture following a closely cut hay or greenchop harvest until 18 to 20 inches of regrowth has occurred. Avoid pasturing these crops after a killing frost for three to five days or until all growth has turned brown. Apply phosphorous and potassium fertilizers according to soil tests and avoid excessive application of nitrogen fertilizer. High rates of nitrogen fertilizer combined with low soil levels of phosphorous can double the prussic acid content of young leaves. If in doubt, harvest the crop as hay or silage. Prussic acid dissipates as the forage lies in the windrow and studies have shown that the ensiling process reduces the prussic acid level by 50% to 80%.

While all the sorghum forages have the potential to produce dangerously high levels of prussic acid, there are differences among species and cultivars. A ranking of these products from lowest potential to greatest is as follows: (1) varietal sudangrass, (2) hybrid sudangrass, (3) hybrid sorghum/sudangrass and (4) forage sorghums.

NITRATE POISONING

The nitrate content of sorghum and pearl millet forages can be high, depending upon growing conditions and the source and amount of nitrogen fertilizer. Excessive rates of nitrogen fertilizer from any source can result in a problem. However, nitrate accumulation by these forage crops is usually greater when nitrate fertilizers are used rather than ammonium sulfate or urea. Low light intensity, frost, hail, low temperature, disease infection, and insect feeding can result in high nitrate content in these forages.

High nitrate forage consumed by livestock results in poor growth, reproduction problems and poor milk production. If nitrate content of forage is very high, rapid death can and does occur. Nitrate is converted by the animal into nitrite. Nitrite absorbed into the blood converts normal hemoglobin into methemoglobin, which is incapable of transporting oxygen to body tissues.

Treatment for animals suffering from nitrate toxicity usually consists of an intravenous injection of methylene blue solution, which brings about the reconversion of methemoglobin to hemoglobin. If nitrate problems are suspected, a representative sample of the forage should be sent to a laboratory for analysis. If the suspect forage is confirmed to be dangerously high in nitrate, it should be mixed with other roughage before feeding or discarded.

The risk of nitrate poisoning can be substantially reduced by sound management practices. Avoid grazing or harvesting these crops immediately after periods of drought, hail, or frost. Do not make excessive nitrogen fertilizer applications. It is best to apply a portion of nitrogen fertilizer preplant (50 to 75 pounds) and then after each grazing period or harvest, make relatively small applications (30 to 40 pounds). Avoid grazing these products after harvest as hay or greenchop until sufficient regrowth has occurred (nitrate accumulates in lower parts of the stalk). Harvest in the afternoon or evening, after the crop has been exposed to several hours of sunlight. Avoid harvesting on cloudy, overcast days. If forage is suspect, harvest as silage. The ensiling process reduces nitrate concentration by 40% to 60%. Finally, never allow hungry animals to pasture potentially high nitrate crops without first feeding roughage to reduce initial intake.

Generally, the sorghum sudangrass hybrids and pearl millets accumulate higher levels of nitrate than does sudangrass. While it is difficult to define the level of nitrate that will produce toxicity symptoms or adversely affect animal performance, when the line is crossed the results may be devastating.

CHAPTER FOUR: CEREAL SMALL GRAINS

The cereal small grains – wheat, oats, barley, and rye offer Kansas livestock producers great versatility in forage systems. Regardless of the species utilized, these crops meet or exceed nutritional requirements of grazing livestock. Small grain forage is high in protein and low in fiber during most of the winter grazing season. Crude protein, on a dry matter basis, normally ranges from 15% to 34%. Properly managed, the small grains may be utilized for pasture until the onset of jointing with little or no reduction in grain yield.

Weight gains of 1.25 to 1.5 pounds per day per animal are usually obtained when grazing cereal small grains during the fall, winter, and spring months.

Small grains may be seeded in pure stands, in mixtures, or over-seeded or sod-sown in perennial grasses. Likewise, small grains may be utilized in the production of high quality hay or silage for beef and dairy operations.
All the cereal forages are annuals, although some are fall sown and others are spring sown. They tiller from nodes at the base of the plants and thus, a single plant usually is composed of numerous tillers. In early stages of development, these tillers are primarily leaf, and become stemmy as maturity approaches. At jointing (stem elongation), a bud is formed just above the first nod. If the bud is damaged by grazing livestock, grain production will not occur. Grazing livestock must be removed at or preferably before the onset of jointing if a grain crop is desired.

When utilizing small grains as a dual-purpose crop, i.e., as a forage and grain crop, the seeding date should be 2 to 4 weeks earlier than for grain production alone. Likewise, the seeding rate should be increased 30% to 50%.

The principal fertilizer requirement of small grains is nitrogen. If small grains are to be utilized for forage, a fall application of nitrogen fertilizer is critical. However, excessive rates of nitrogen may result in luxury consumption and high forage yields cannot be maintained without additional nitrogen later in the season. Normally, small grains are fertilized with nitrogen, phosphorous and potassium at planting time and topdressed with additional nitrogen in winter and sometimes spring. If a grain crop is contemplated, it is especially important to split nitrogen applications.

There are a couple of useful "rules-of-thumb" to aid in determining nitrogen topdress needs: (1) It takes approximately 2 pounds of nitrogen to make one bushel of grain yield, and (2) One pound of beef is produced from 8 to 10 pounds of dry forage, which contains approximately three-tenths (0.3) of a pound nitrogen. A simple and fairly accurate method of determining the amount of nitrogen removed by grazing livestock is to estimate the pounds of beef produced per acre and multiply that value by 0.3.

EXAMPLE:

Over the course of the fall and winter seasons, the gain of grazing cattle is estimated at 100 pounds per acre forage.

$100 \times 0.3 = 30$ pounds nitrogen removed

From a previous soil test the amount of residual nitrogen in the soil is determined, add this value to the amount of nitrogen applied preplant and then subtract the estimated amount removed by grazing livestock. Bearing in mind that it takes roughly 2 pounds of nitrogen to produce 1 bushel of grain yield, topdress nitrogen needs can be determined.

It is fairly common practice in Oklahoma to sow small grains into bermudagrass sod. This management practice also holds potential for southeastern and east-central Kansas. Bermudagrass pastures are effectively out of production for five months a year (November through April). With a sod-sowing system, a small grain crop can be grown during this period with little overlap. The bermudagrass should be grubbed down or clipped prior to attempting to establish small grains. Fertilizer is often banded with the seed as it is nearly impossible to incorporate phosphate and potash without severely damaging the sod. However, banded nitrogen and potash must not exceed 30 pounds total per acre or severe injury to small grain seedlings may result.

Sod-sown small grains provide a much more efficient utilization of land, high quality winter and/or spring grazing, and potentially a grain crop. Even though both crops will be growing simultaneously in the spring, the small grain is usually far enough ahead of the bermudagrass that a satisfactory grain yield can be achieved in spite of the competition. The determination of whether or not to attempt a grain crop can be delayed until the first of March. Harvest of a grain crop is relatively easy, as the bermudagrass sod provides excellent footing even in wet conditions.

When utilizing small grains for hay or silage, the most important management consideration is timeliness of harvest. Small grain forage should be cut in the boot-to-early-head stage. At this stage of development, small grain forage is very good in quality. Greater digestibility, higher protein, and greater energy of small grain forage cut in the boot-to-early-head stage, more than offsets the greater dry matter yield of forage cut in later stages of maturity.

In Kansas, small grain forage is most often utilized for pasture. It is generally best to allow forage to accumulate prior to initiating grazing. Six-to-ten inches of accumulated growth will usually result in excellent full-season production provided other factors are in balance, (that is moisture, fertility, temperature, etc.) On the more prostrate winter wheat varieties 4 to 8 inches accumulation is acceptable.

Clipping studies have shown that the more frequently small grains are harvested, the lower final forage yield. Continuous grazing of small grains is the least productive management system employed by cattlemen. Rotational grazing with periods of grazing deferment of 20 to 40 days, depending upon the season and growing conditions, results in dramatic improvement of forage yield and animal performance. Forage should be grazed to a minimum height of 2 to 4 inches. Shorter grazing height will generally reduce final yield. The only exception is in the last harvest of a graze-out situation. At the end, "take it all!" although, agronomically speaking, about 1 ton of plant material should be left for maintenance of soil structure and erosion control.

In a rotational grazing system, generally the more pastures the better; two are better than one, three are better than two, four are better than three and so-on. Typically, the Kansas cattleman gets bogged down by the requirements of fencing and watering involved in rotational grazing systems. While this is a valid consideration, these problems are relatively easily overcome. A single strand of electric wire is sufficient to segregate cattle from one pasture to another. Below is just one example of a rotational grazing approach.

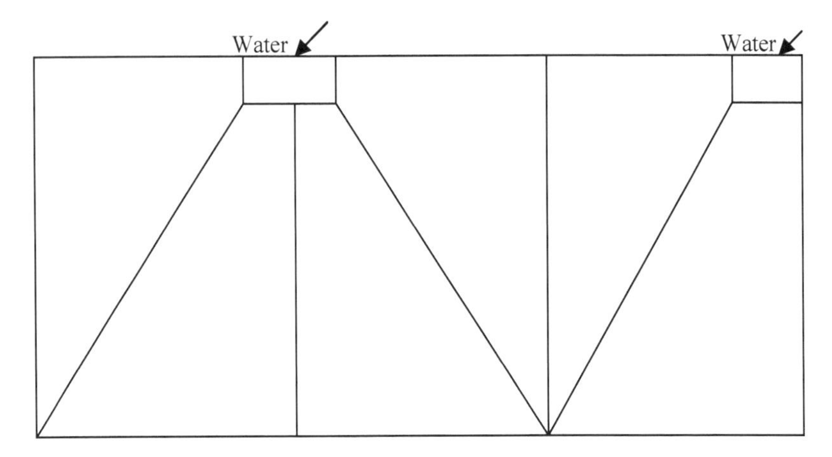

It is best to keep the area of water and dry roughage supplementation out of the small grain field if possible.

Rotational grazing of small grains increases forage production, allows greater stocking rates, increases beef production per acre, and generally increases the profitability of the enterprises.

There are problems and serious management considerations in utilizing small grain forage that must be addressed by the producer/manager. Many soils utilized for wheat grain production in southeastern Kansas consist of a shallow layer of top soil under which lies a very heavy "B" horizon. Cattle can cause severe compaction when such soils are wet. Likewise, cattle can introduce new weed species to cropland via their digestive systems and/or tracking.

One must remember that if a grain crop is desired, livestock must be removed from small grain pasture before the onset of jointing. Good gains are produced from small grain pasture during January and February only if sufficient forage is available. Very little growth is produced at this time and, unless forage is stockpiled in the fall/early winter, there will not be sufficient forage to maintain animal performance. Overgrazing and grazing through stressful periods can significantly lower subsequent grain yields.

There is an extreme danger from tetany when small grain forage is grazed by cows about to calve or those that have recently calved. This is not conventional grass tetany. Calcium rather than magnesium is apparently the primary mineral deficiency.

Small grain forage is rapidly digested, and the potential of bloat is another problem that must be dealt with by the producer/manager. The danger of bloat, while not eliminated, is minimized by properly managed rotational grazing and the provision of adequate amounts of dry roughage and anti-foaming agents (Bloat Guard).

Haying or harvesting small grains for silage must occur at the optimum stage of maturity (boot-to-early-head), which often conflicts with planting of spring-sown crops; that is grain sorghum and soybeans, or the first cutting of other grass or legume hay crops.

Small grains, carefully managed, make an excellent forage. They vary in seasonal distribution of yield such that sowing mixtures of two or more small grains can be advantageous. Of course, sowing small grain mixtures precludes the opportunity of harvesting a pure grain crop.

In selecting species and cultivars of small grains for utilization as forage, the producer/manager should consider some basic criteria: (1) Total forage yield is of utmost importance as total livestock gain will ultimately be determined by the amount of forage produced. (2)

Seasonal distribution of yield should be managed to minimize the need for supplementation. (3) Many of the hard red winter wheat cultivars commonly planted in Kansas have a semi-prostrate growth in the fall; those with more upright growth provide forage more readily available to grazing livestock. (4) Insects such as greenbugs, Hessian fly, etc. and diseases such as crown and root rots, soil-borne mosaic virus, etc. can substantially lower forage yield and quality. Cultivars with genetic resistance to these pests are often the best forage producers. (5) Wheat coleoptiles become shorter when soil temperatures are high and some varieties do not germinate well in hot soils. Therefore, rye or triticale cultivars are often a better option when planting in August or early September.

Small grain forages offer great versatility to the Kansas livestock industry. They provide supplemental nutrients for cow-calf operations, make a significant contribution to the stocker cattle industry and are being utilized more and more in finishing programs. With the shifting of consumer preference towards leaner red meats, small grain forage will continue to gain in popularity and usage.

WINTER WHEAT

Kansas is recognized around the world as a major production area of hard red winter wheat. It traditionally leads the United States in acres planted and bushels harvested. Currently, almost 100 varieties of winter wheat are marketed in the state. Many of these cultivars are well suited to forage systems as well as for grain production.

Winter wheat produces lush, palatable, nutritious forage high in protein and minerals. Cattle grazing wheat pasture can average as much as 1.5 to 2 pounds daily gain throughout the fall, winter, and spring grazing periods.

Carrying capacity and length of grazing season can and does vary greatly from year to year, depending upon precipitation, average temperature, disease pressure, (soil-borne mosaic virus, root and crown rots, etc.) insect pressure, (greenbugs, Russian wheat aphid, Hessian fly, etc.) and the grazing system(s) employed. Wheat should be firmly anchored by secondary roots before allowing livestock to pasture. Overgrazing can substantially lower subsequent grain yields, and in central and western Kansas, often results in wind erosion. Allowing livestock to graze when fields are muddy can destroy wheat stands and on shallow soils can cause compaction. Livestock must be removed in the spring prior to jointing (stem elongation) if a grain crop is desired.

Seeding rates of winter wheat utilized for forage range from 90 to 120 pounds per acre on dryland production in eastern Kansas and under irrigation in other areas of the state. In western Kansas, seeding rates range from 45 to 75 pounds per acre on dryland production. Date of planting should be two-to-four weeks earlier than normal seeding dates used for grain production.

In recent years, wheat silage as a feed for dairy and beef cattle has gained in popularity. The extent to which wheat is utilized as silage has fluctuated, depending upon the cash grain price and government programs. The best time to harvest wheat for silage is from the boot-to-early-head stages of growth. In Kansas, this time frame for cutting is often not more than 7 to 14 days. Thus, it is imperative to begin harvest promptly when wheat is in the boot stage in order to finish by the early-head stage. While greater forage yields may have been obtained by harvesting at later stages of maturity, resulting forage will have less energy and protein.

Winter wheat grain has long been recognized as an excellent feed for swine. However, it has only been in the last 15 to 20 years that it has been widely utilized in finish cattle rations. When wheat grain prices are low in comparison to corn and sorghum, usage dramatically increases. Wheat is generally higher in protein than other feed grains and is especially good as a supplement in high roughage rations. Rolling wheat is a satisfactory method of processing for feed; wheat must not be ground too finely. Feeding wheat grain requires more care than the feeding of corn or sorghum. If an animal consumes too much wheat, death may result from the rapid fermentation of the grain in the forestomach. Therefore, wheat should be limited to no more than 40% of the ration and care should be taken to ensure the ration is completely mixed. It is best to gradually increase wheat grain intake over a 35 to 40 day period before putting cattle on "full feed." Wheat grain should never be used in creep-feeding calf rations.

Variety selection is often a critical management decision when utilizing winter wheat for forage. In western Kansas, those varieties with resistance to the Hessian fly and tolerance to wheat streak mosaic virus will be more consistent in performance at the earlier planting dates. In central and eastern Kansas, tolerance to soil-borne mosaic virus, spindle streak mosaic virus, leaf diseases, and the Hessian fly can often mean the difference between a profitable forage/grain system and a mediocre or poor return on the investment. Varieties differ in their ability to provide rapid ground cover in the fall. Likewise, some varieties grow almost flat against the ground, while others offer grazing livestock an easily accessible, upright forage growth. When utilized for hay or silage, the taller, later-maturing varieties may produce greater forage yields. However, in central and eastern Kansas, these later-maturing varieties are generally subjected to greater pressure from the leaf diseases (such as leaf rust) that may substantially lower yield and quality.

OATS

In Kansas, spring oat cultivars are usually recommended rather than winter types. Even in the southern-most counties of the state, the cold tolerance and winter hardiness of the winter oat varieties currently available is questionable.

When mature, oat plants normally range from 3 1/2 to 5 1/2 feet tall. They have narrow, blade-like leaves 6 to 9 inches in length. The head is a many-branched panicle and each branch (spikelet) bears two self-fertilized flowers. The fruit or grain produced by each flower is a caryopsis, a single-seeded fruit in which the hull is firmly attached to the seed.

While oats are frequently grown as a feed grain elsewhere, in Kansas, they are most often utilized as a forage crop for grazing, hay, and silage. They produce a lush, high protein forage growth with better palatability than other small grains in the spring. Oats are often the preferred companion species for new plantings of legumes and grasses. When utilized as a companion crop, the oat seeding rate should be reduced by 1/2 normal rates so that competition for sunlight, moisture, and nutrients is limited. The oats are harvested as hay, silage or grain the summer of establishment and the grass or legume is harvested the following season. The oat seedlings serve to protect the fragile grass and legume seedlings from blowing soil particles, excessive temperatures, and hot, drying winds, and they also aid in weed suppression. It is often advantageous to harvest the companion oats in an early stage of maturity as hay or silage in order to minimize competition with the grass or legume planting.

Oats managed properly and harvested in a timely manner can be made into satisfactory silage. They lose much of their milk producing potential upon reaching the milk stage of grain-fill. While forage yield may be greater at later stages of maturity, the greater digestibility and quality of oat forage harvested in the boot-to-flowering stage will more than offset the yield difference. Oats harvested in these early stages of development should be wilted to less than 70% moisture, and can be preserved in tight conventional silos at 60% to 65% moisture. Harvest in the hard dough stage results in lower moisture and the forage will not pack well. Likewise, as oat plants mature, the stems hollow out, trapping air that usually results in molding of silage.

Oat hay, like silage, varies in quality depending upon the stage of maturity at harvest. While hay yield will increase up until the milk stage, hay harvested in the boot-to-flowering stage will be superior. Oats grown for hay are frequently fertilized with 70 to 90 pounds nitrogen per acre. To maximize forage quality and minimize lodging problems, nitrogen application should be split with a portion applied preplant and the balance top-dressed as stems begin to elongate. Oat hay is considered excellent in palatability and quality and is in demand as dry stock dairy and horse feed.

When selecting cultivars to use in forage systems, greater production usually results from the "forage-type" cultivars as opposed to the "grain types" (Larry, Bates, Ogle, etc.). The two most commonly planted forage-type spring oat cultivars in Kansas are Jerry and Troy. If planting winter oat cultivars for forage production in Kansas, it is generally best to utilize them in combination with or mixtures of various other small grains and/or legumes. Cimarron, Bob and Nora are the winter cultivars most widely used in southern Kansas.

When oats are grown as a feed grain, the Kansas producer anticipates high grain yields, much of which is used on the farm. With the legalization of horse racing in Kansas, a potentially profitable cottage industry of oat grain and hay production for race tracks and horse owners will likely develop. Bear in mind, the horse industry demands bright, sound, high quality oats heavy in test weight, and white grain color is greatly preferred to either yellow or brown. Likewise, with the recent public awareness of health aspects of oat bran, demand for oat grain for human consumption is expected to escalate.

BARLEY

In Kansas, both spring and winter types of barley are grown. Winter barely is much less hardy than winter wheat and does not extend as far northward as does the production of winter wheat. While there is a great deal of overlap, generally in southern Kansas the winter cultivars such as Post, Schuyler, and Kanby are recommended, while in northern Kansas the spring cultivars such as Robust, Morex, and Hazen are often more productive.

Barley is tolerant of mildly alkali and acidic soil conditions, drought, and frost. However, it produces best where the spring season is long and cool and on deep loam to clay soils with adequate fertility and near neutral pH. Kansas, due to its geography and climate, cannot consistently produce good quality malting barley and as a result, barley is almost exclusively grown as a feed grain. Barley grain is hard and should be rolled or cracked before feeding to livestock. Good quality barley (test weight greater than 45 pounds) is considered to be 95% the feeding value of corn.

Winter barley is often utilized for pasture during fall, winter, and spring. It produces a lush, palatable forage rapidly in the fall and is considered superior to wheat and rye in autumn and early winter months, but wheat and rye are more productive forage producers in late winter and spring. Barley awns may have sharp barbs that are longer and sharper at their base and narrow and blunt towards the tip. If fed or grazed by livestock, the rough awns may cause mouth and digestive tract problems. However, there is great variation in the form of hoods and awns. Some varieties are almost completely awnless, while others possess very prominent awns. The hulls adhere tightly to the seed in most commercial varieties, but hull-less or naked barleys also exist.

Seeding rates vary from 1 1/2 to 2 bushels per acre in eastern Kansas and under irrigation in other areas of the state to 30 to 48 pounds dryland in western Kansas. If barley is to be utilized for pasture, the seeding rate is usually increased by 30% to 50%. Date of seeding should be two to four weeks earlier than for grain production. Seedbed preparation and fertility requirements are similar to those of winter wheat. Barley can be made into good quality hay or silage if harvested in a timely manner (boot-to-early-head stages), and put-up at the correct moisture.

Barley is subject to attack from several diseases and insects including smut, rust, barley yellow dwarf virus, armyworms, chinch bugs, and greenbugs. Both chinch bugs and greenbugs seem to prefer barley to any other host small grain. Thus, the producer/manager must "scout" barley fields periodically to determine the presence and/or extent of infestation.

In Kansas, barley acreage is largely dependent upon the federal farm programs, and has fluctuated accordingly. When selecting a variety to plant, consideration should be given to straw strength, as lodging is perhaps the most serious production problem.

WINTER RYE

Winter rye is excellent forage in Kansas. It tolerates acid soils and wide variations in climatic conditions. While rye thrives on fertile, productive soils, it also performs very well on marginal and/or sandy soils. Winter rye is hardier than the other winter small grains. Rye grows at cooler temperatures and thus provides greater forage yield in the fall and winter than wheat.

The rye plant generally obtains a height of 5 to 7 feet. The leaves are larger and coarser than those of wheat and are bluish in color. The stems are topped by slender spikes (flower heads) that are longer than those of other cereal small grain species and are more heavily bearded.

Planting dates of winter rye utilized for forage range from late August thru September. Seeding rates vary from 75 to 100 pounds per acre on dryland in eastern Kansas and under irrigation in other areas of the state to 40 to 60 pounds per acre dryland in western Kansas. Cultural and management practices are similar to those used in winter wheat production. The most important management considerations are as follows:

1. Delay initial grazing to allow forage accumulation of 6 to 10 inches.
2. Strictly maintain a minimum harvest height of 2 to 4 inches, until the last harvest.
3. Use rotational grazing with deferments of 20 to 40 days, depending upon the season and growing conditions.

Rotational grazing will result in the best utilization of available forage and maximum animal performance. Continuous grazing, on the other hand, leads to more pronounced spot grazing, larger waste due to trampling and fouling by excreta and generally lower animal performance.

Manage for hay or silage as for winter wheat, that is, harvest at the proper stage of maturity and put-up at the proper plant moisture.

While winter rye is generally superior to other small grains in fall and winter forage production, growth terminates rather rapidly in May. Winter wheat is a more vigorous forage producer in the spring and provides at least an additional two weeks grazing. These differences in seasonal distribution of yield lend credibility to the planting of blends. A mixture of 60% to 70% winter rye and 30% to 40% winter wheat results in maximum season-long forage production. The inclusion of a winter annual legume such as winter peas or hairy vetch can increase yields and make a better quality, more balanced feed.

In Kansas, the majority of winter rye planted is "variety not stated" of Nebraska, South Dakota or Minnesota origin. However, the southern varieties (Elbon, Bonel, etc.) offer superior forage yield.

TRITICALE

Triticale is a synthetic hybrid resulting from crossing wheat and rye. The wheat plant is used as the female parent and rye is the pollen donor. The resulting hybrid is sterile, but if treated with colchicine to induce polyploidy, fertility may be restored. Triticale usually combines the yield potential and grain quality of wheat with the disease and stress tolerance of rye.

In Kansas, triticale is used almost exclusively as a forage crop. However, there is much interest in utilizing triticale as a feed grain. The starch contained in triticale grain is readily digested and the protein content is superior to wheat.

A firm, well-prepared seedbed generally results in uniform emergence and vigorous seedlings. No-till seeding into crop residues is a practice gaining in popularity, but care must be exercised to insure proper seed placement and good seed-to-soil contact is achieved.

In eastern Kansas and irrigated production in the west, seeding rates of 90 to 100 pounds per acre are recommended. In western Kansas dryland production, recommended seeding rates range from 50 to 90 pounds per acre (depending upon annual precipitation and soil type). Producers should plant seed at a depth of 3/4 to 1 1/2 inch in late August or September. Uniform seed depth is critical for optimal stand and yield.

Soil fertility management begins with a soil test. With fertilizer costs where they are today, a forage producer must have an accurate basis upon which to determine crop needs. Whether grazing, haying, or ensiling, it is best to split nitrogen applications. A 2-ton-per-acre crop with 14% crude protein will remove approximately 90 pounds of nitrogen per acre. A forage yield of 4-ton-per-acre would be double (approximately 180 pounds of nitrogen). No more than 1/2 of the required nitrogen should be applied at planting. The balance should be top-dressed in one or two operations in late winter and/or early spring. Too much fall applied nitrogen produces excess growth that tends to mat and deteriorate in quality. It also results in greater winter-kill and may promote disease issues. Apply phosphorus and potassium according to the recommendations of a soil test.

As a hay or silage crop, triticale should be harvested in the boot stage (stage 9) prior to heading. In Kansas, it generally yields 2 to 4 ton of dry matter per acre. Triticale may be grazed in the fall and spring, ensiled, or baled. When grazed, triticale is nearly equal to rye for fall and winter pasture and superior to rye in the spring. When compared to wheat, triticale is much better in the fall and winter and just as good in the spring. Rotational grazing will result in the best utilization of available forage and maximize animal performance.

Manage for hay and silage as for wheat; that is, harvest at boot stage and put-up at the proper plant moisture. It is important to bale or ensile in a timely manner, as nutritive value will decline if left in the field too long.

Including a winter legume such as winter peas or hairy vetch will increase forage production and provide additional nutrients.

In Kansas, there are many varieties available to plant. The northern varieties (Pika) are recommended due to their superior winter hardiness. However, it is often difficult to obtain these varieties because triticale harvest in the Dakotas and Canada coincides with our planting window.

CHAPTER FIVE: LEGUMES

Legumes are amazing plants. They are unique in their ability to gather nitrogen from the atmosphere and convert it to a form beneficial to the plant. The legume achieves this through the activity of bacterial colonies, (nodules), which form on the roots when the proper strain of rhizobia bacteria is present in sufficient number in the soil upon germination of the legume seed. Some legumes are very efficient at gathering nitrogen and convert enough to supply their own needs as well as those of surrounding plants.

All legumes bear fruit in the form of a pod containing seeds. These plants may be annuals, perennials, or biennials, and include some shrubs and trees. Legumes also include a great number of ornamental plants and plants valued as sources of such products as dyes, drugs, resins, perfumes and wood. They also include some of the most important of all food and forage crops, such as clovers, peas, beans, alfalfa, peanuts, and soybeans. Many legumes are exceptionally rich in proteins and are practically the only non-meat source of some of the amino acids essential to the human diet. Unfortunately, not all legumes are beneficial. Some can be aggressive weeds (such as Kudzu and Sericea lespedeza) and many are poisonous to grazing livestock (loco weed, crolateria, etc.).

Before low-priced nitrogen fertilizer greatly reduced their usage, many legumes were utilized as green manure crops. Increased fertilizer prices, along with recent concern and awareness of ground water nitrate levels and stream pollution due to erosion runoff, may result in tremendous demand for legume crop seeds in the coming years.

Many legumes are utilized in pure stands or in combination with grasses for grazing and/or hay production. When utilized in combination with grasses, legumes increase forage yield and protein content, improve palatability and digestibility of grassland forage, supply nitrogen, improve soil structure, and can substantially increase carrying capacity of pastureland. However, if legumes are to co-exist with grass and be productive, the grass/legume mixture requires good soils with good to excellent drainage and relatively high pH values. While legumes provide their own nitrogen, they are generally heavy users of other nutrients, in particular potassium and phosphorous.

All legumes should be inoculated before planting. Inoculation is the practice of introducing commercially prepared rhizobia bacteria into the soil. This may be accomplished by applying the proper amount and strain of bacteria to the seed prior to planting or by metering the inoculant into the furrow at planting. While most soils contain rhizobia bacteria, the strain may be poorly suited to the legume in question (improper strains can be lazy or parasitic to the host crop). Even in those instances where specific legume crops, such as soybeans, are grown continuously on a particular field, it's a good practice to inoculate. The population of rhizobia bacteria in any given soil is strongly influenced by environmental factors. Heat, dryness, low organic matter, and acidity are detrimental to the survival of rhizobia bacterium. Producers are definitely not penalized by an overabundance of rhizobia in the soil. On the other hand, if populations are inadequate, legumes may starve from a lack of nitrogen.

ALFALFA

Alfalfa is grown over a wide range of soil and climate conditions and has an important role in crop rotation through its positive effects on soil fertility, soil structure and soil erosion control. It is a versatile crop, utilized in Kansas for pasture, hay, silage, greenchop, green manure, and as a cash crop of both hay and seed.

Alfalfa is a herbaceous plant with trifoliolate leaves arranged alternately on the stem. A perennial with a deep tap root system, alfalfa has no peer as a soil improvement crop. A mature plant may have from 5 to 50 stems that vary in length from 1 to 3 feet, depending upon growing conditions. At maturity, stems bear 1 to 10 flower clusters called racemes. The flowers are usually some shade of purple although white, yellow and mixtures of colors can occur. Generally there are from 5 to 25 flowers per raceme and each successfully pollinated flower produces a coiled pod that contains up to ten seeds. The seeds are kidney-shaped and vary from bright yellow to olive green to light brown. Alfalfa is highly cross-pollinated, usually requiring bees or other insects to "trip" flowers, allowing the stigma to come into contact with pollen on the insect's body.

Establishment

Choose the best soil possible to produce alfalfa. Deep, well-drained loam or clay-loam soils are best. These soils possess the necessary characteristics for top production, which is good water-holding capacity and adequate infiltration rates. Alfalfa can be produced on less than ideal soils, but forage yield will suffer. Heavy clay soils are slowly permeable and tend to water-log. Shallow soils reduce the root zone and water-holding capacity.

The producer must plan ahead to successfully produce alfalfa. Soil pH should be near neutral and potassium and phosphate levels in the soil should be in the high range. In Kansas, a late summer (August-September) planting date is best. Later fall plantings may be attempted but run the risk of being killed by cold weather before seedlings become firmly established. Spring plantings of alfalfa are often successful; however, they entail greater risk of stand failure. Early spring planting dates run the risk of a late freeze, while late spring plantings may not allow the alfalfa plants enough time to establish an adequate root system before the onset of hot/dry weather. If spring plantings are attempted, usually mid-March to mid-April planting dates are best.

Preparation of the seedbed is one of the most important factors in the successful establishment of alfalfa. The surface of the seedbed should be fine (free of clods) and very firm. Below the surface, soil must be loose enough to allow seedling roots easy penetration. The surface must not be powdery; if conditions force the field to be worked more than what would be ideal, it is best to wait for a rain to settle and firm the surface.

Moisture at the surface of the seedbed is not essential. In fact, successful alfalfa plantings are often made when the surface of the soil is dry. It is helpful when there is good moisture 1 to 3 inches below the surface.

Alfalfa must be planted shallow. Planting depths of 1/4 to 1/2 inch are best. Generally, 12 to 18 pounds of seed per acre is required to insure good stands. Higher seeding rates are needed in the eastern part of the state than in the western part.

Use and Management

Alfalfa is considered the "Queen of the Forages" for good reason. At 10% bloom stage, alfalfa contains approximately 20% crude protein and total digestible nutrients range from 50% to 60%.

If high production is the goal in beef and dairy operations, alfalfa should be an essential ingredient of the feed rations. Alfalfa hay also serves as roughage, necessary to keep the rumen functioning and to keep milk fat from dropping.

Alfalfa pastures are commonly used for swine and horses. Being mono-gastric, the danger of bloat is eliminated with these classes of livestock. Bloat of ruminant animals is always a danger. Daily administration of an anti-foaming agent at labeled rates will control bloat. Improper grazing management can severely thin and shorten the life of an alfalfa stand.

Production of high quality maximum hay yields requires considerable management skill. The following guidelines are key in successful alfalfa hay production:

1. Establish a thick stand. Plant 12 to 16 pounds per acre. Half the battle with alfalfa is getting a good, weed-free stand, so plant only high quality seed and direct seed using no nurse crop.
2. Use superior yielding, multiple pest-resistant varieties. The higher cost of the seed will be returned many times through the life of the stand.
3. Soil pH should be 6.7 or above, and soil should be a deep loam or clay loam for maximum production.
4. Soil drainage should be good to excellent.

5. Apply fertilizer the year of establishment. Work in 50 to 60 pounds phosphate and 250 to 300 pounds potash fertilizers per acre prior to sowing and before seedbed preparation.
6. Take the first cutting when plants are in the late-bud stage of maturity, (in Kansas this is usually late May); the second, third and fourth cuttings should be removed when the first flowers appear. The last cutting should be delayed until after October 15th, to allow plants to store adequate carbohydrate reserves to maintain the stand through the winter.
7. Topdress with fertilizer annually. Apply 12 pounds phosphate and 60 pounds potassium for each ton removed. Also, apply 2 pounds of boron semiannually.

Varieties

Variety selection is a critical management decision that is often overlooked by Kansas producers. The investment required to establish alfalfa is too high to ignore high quality seed of well-adapted multiple pest resistant varieties. Many Kansas alfalfa producers grow what they call "Kansas common." In reality, there is no longer any true Kansas common strain. Most seed referred to as common are mixtures of commons and old varieties. Some may perform well while others are poorly adapted. Multiple pest-resistant varieties adapted to Kansas result in higher yields, longer stand life, and reduced dependence upon pesticides.

There are many varieties well adapted to Kansas that offer the Kansas farmer the opportunity to maximize production and profits. In southeast Kansas, where most soils have a fairly shallow layer of top soil and a very heavy B horizon, we like the America's Alfalfa varieties. These varieties offer a unique plant type with exceptional traffic tolerance and resistance to Phoma crown rot – the "traffic disease." These unique genetics are the result of more than a decade of parent stock selection to

- handle damage from heavier baling equipment.
- meet the needs of growers who want to cut frequently for higher quality.
- endure harvest during less-than-ideal conditions.
- survive tough winters and late fall harvest.
- start strong at seeding in the presence of wet soil diseases.
- demonstrate stronger first-year yields and less yield fade in later years.

Traffic tolerant varieties feature deep-seated crowns with finer stems and greater leaf mass, and larger roots that store more energy promoting faster yield and recovery. These varieties combine high resistance to all major alfalfa diseases with the best wheel-traffic and grazing tolerance available.

There are many other varieties that consistently perform well above the average in the Kansas Alfalfa Variety Performance Tests. Descriptions of these varieties should be obtained from local dealers.

RED CLOVER

Red clover is the most widely grown of all the true clovers. It is grown alone or in combination with grasses for hay, pasture, and soil improvement. Most American red clovers are the early-flowering type, known collectively as "medium red clover." This type is characterized by producing two or three hay crops per year and having a biennial or short-lived perennial growth habit. Mammoth red clover is the principal late-flowering type grown in the United States. The late-flowering or single-cut type usually produces one hay crop plus an aftermath.

Red clover requires ample precipitation to produce high forage yields; thus, it is limited to the eastern half of Kansas. It performs best on those soils with good water retention and drainage qualities. Deep loams, silt loams, and even fairly heavy soils are preferable to shallow or sandy soils.

Taproots of red clover plants can penetrate depths of up to 6 feet and numerous roots will develop laterally from the taproot, making red clover an excellent soil improvement crop. Usually, the taproot disintegrates in the second year and surviving plants rely upon the development of a strong secondary root system.

Red clover is a herbaceous plant made up of numerous leafy stems arising from a crown. The flowers are borne on heads of compact clusters at the tips of the branches. Each flower cluster can consist of as many as 125 individual flowers. Flower color is rose purple or magenta. The shape is similar to pea flowers, although much smaller. Seeds are contained in coiled pods and are oblong to slightly kidney-shaped and vary in color from pure yellow to purple.

Red clover will grow on moderately acid soil and be moderately productive; however, for maximum production soil pH should be 6.4 or higher.

Establishment

In Kansas, red clover seed is usually broadcast into stands of winter wheat during February or March. Freezing and thawing of the soil results in adequate seed/soil contact. Many producers prefer to broadcast seed on top of snow. The presence of tracks in the snow aid in uniform broadcasting by reducing overlapping and skips. When the snow melts, it effectively pulls seed down into the soil.

If planting red clover with a spring-seeded small grain companion crop, such as oats, it is best to use a drill equipped with legume seed boxes and plant both crops at the same time. Sow 10 to 12 pounds red clover seed per acre and reduce the seeding rate of the small grain by 1/3 to 1/2. The practice of using small grain companion crops has proven highly productive in eastern Kansas. Usually, the small grain is harvested as cash crop in late spring or early summer and the red clover provides grazing or a small hay crop in autumn. The following year provides the opportunity to get three excellent quality hay crops or two hay crops and a seed crop from the red clover.

Red clover is also extensively utilized in pasture mixtures and for renovating old pastures. It is one of the easiest legumes to establish in closely grazed or disturbed sods. It is generally best to disk sod, or disturb by some other means, in late fall or winter and sow the clover seed in February. Usually, 4 to 6 pounds of good quality seed is adequate to provide 20% to 30% legume in the grass stand.

In southeastern Kansas, many producers sow red clover in late summer rather than winter. While this is considered to be risky from the standpoint of potential stand failure, late summer plantings can result in much greater production the first season. If sowing in late summer, adequate moisture is imperative and seedlings must be firmly established (secondary roots present) before the onset of winter.

Use and Management

In the first year of the stand, one hay crop may be attempted, weather permitting. If the newly seeded red clover blooms by September 5 in northeast Kansas or by September 15 in southeast Kansas, the growth should be removed by grazing, haying, or clipping. Red clover allowed to bloom in the seedling year has much less winter-hardiness than plants that do not bloom. If a hay crop is harvested the seedling year, the stand should not be grazed prior to freezedown. In succeeding years, the first hay crop should be taken in the pre-bloom to early-bloom stage. Subsequent hay crops should be taken when plants are in early-bloom stage. Often, the first crop is harvested as hay and subsequent crops harvested by grazing animals. Red clover as pasture should be grazed at six to seven week intervals utilizing rotational rather than continuous grazing. Bloat can be a serious problem with cattle grazing red clover. Grazing animals should be observed closely and provided with plenty of dry roughage. Anti-foaming products (Bloat Guard) are recommended but care must be taken to insure each animal regularly consumes an adequate supply.

Red clover is routinely over-seeded into pastures and meadows to improve quality and yield. In a study conducted by Kansas State University extension researchers on a farm near Girard, red clover interseeded into tall fescue dramatically increased the amount and quality of forage produced. Red clover was interseeded at a rate of 12 pounds per acre in February, 1981, into random plots of tall fescue. The interseeded plots received 0-40-40 or 0-80-80 (nitrogen, phosphate and potassium respectively), while the plots left in pure tall fescue received 80-40-40. Two cuttings were obtained in 1982. The first cutting harvested June 10th resulted in the interseeded plots averaging 4.7 tons per acre of forage, 10% greater than the pure grass plots fertilized with 80 pounds. of nitrogen. The second cutting, on August 25th, contained mostly red clover from the interseeded plot and as a result produced 1 2/3 tons per acre, more than twice the forage produced by the pure tall fescue plots.

Varieties

Most red clover seed planted in Kansas is either locally grown or imported from Oregon, and is loosely termed "common" medium red clover. This seed generally performs very well and is lower in cost than improved varieties. However, stand life of common medium red is generally limited to two years, while stands of the improved varieties like Cinnamon Plus often persist three or more years.

CRIMSON CLOVER

Crimson clover is an annual clover often utilized in pasture or haying operations. The leaves and stems of crimson clover resemble those of red clover, but the leaves are rounder and hairier. Seedlings grow rapidly from the crown forming a rosette. In the spring (as the weather warms) flower stems develop rapidly and terminate their growth with long, pointed, crimson-colored flowers.

Crimson clover will grow and thrive on poorer soils than most other clovers. It does not do well in extreme cold or heat and prefers a pH range of 6 to7. Its best adaptation is in eastern Kansas, but it will be successful in central Kansas in all but the driest years.

Establishment

As with all legumes, it is recommended that seed be inoculated prior to planting. However, if mixed with grass for pasture or hay, it is probably not necessary.
Soils should contain medium to high levels of phosphorus and potash at planting, but nitrogen applications are not recommended prior to establishment. Crimson clover may be planted late summer/early fall or late winter/early spring. In Kansas, it is usually sown in late summer, allowing it to maximize forage production. If sowing into existing grass pastures, broadcasting seed in mid-winter when the ground is frozen often results in better stands than drilling seed. When broadcasting or overseeding pastures, 5 to 10 pounds of seed per acre is adequate. If planting crimson clover in a pure stand, 10 to 15 pounds per acre should be sufficient. Seed should be planted at a depth of 1/4 inch. Planting depths of greater than 1/2 inch will result in poor emergence and stands.

Use and Management

Crimson clover may be planted in pure stands as a hay crop, cover crop, or as a green manure crop. Most often in Kansas, crimson clover is sown into existing stands of grass, where it improves forage quality, carrying capacity and forage yield. When utilized in legume/grass pastures, management should be based upon the grass component of the mix. Crimson clover requires a high level of phosphorus but avoid over-fertilizing with nitrogen. High levels of nitrogen fertilizer will weaken clover and other legumes by making them lazy (less vigorous) and less able to compete with the grass component of the mix. When haying, best yield and quality will be achieved if cut in the early-bloom stage.

When used as winter pasture, it is best to utilize a controlled or rotational grazing system that prevents livestock from grazing too closely. Crimson clover grazed too hard in the fall and winter will suffer stand loss and poor forage yield the following spring.

Varieties

In Kansas, the most commonly recommended variety of crimson clover is Dixie.

ALSIKE CLOVER

Alsike clover is a short-lived perennial with a growth habit similar to red clover. It can be distinguished from red clover by the absence of crescent-shaped marks on the leaflets, and its more prostrate growth. Alsike clover can obtain plant heights of 2 to 4 feet; however, the plants tend to lodge unless companion plants (grasses) prop or hold the stems upright.

Alsike clover is adapted to a wide range of soil types and performs on acid and/or wet soils where other clovers will not survive. It is also more tolerant to alkali soil conditions than most clovers. Alsike clover may be utilized in areas of Kansas that receive 18 inches or more annual precipitation. It does not tolerate drought, but will withstand flooding up to six weeks.

Establishment

Due to its small seed size (680,000 seeds per lb) alsike clover is fairly economical to plant. When overseeding pastures, 1 to 2 pounds per acre is usually sufficient to achieve desired legume populations. If seed is drilled, care must be taken not to incorporate seed too deeply. The proper planting depth for alsike clover is 1/8 to 1/4 inch. In eastern Kansas, frost seeding or early spring planting dates are preferred. In central Kansas, early spring (March/April) planting dates are recommended. Alsike clover needs about six weeks of good growing conditions prior to a hard frost or the onset of hot conditions to fully establish.

Use and Management

In Kansas, alsike clover is almost never recommended in pure stands, as there are generally more productive options. However, alsike clover does have some unique applications that make it worthy of consideration. Alsike clover will persist on water-logged soils, making it an ideal companion to grasses on creek and river bottoms. While it is not shade tolerant, it persists in pecan or walnut groves much better than other clovers. Alsike clover also tolerates lower pH than other clovers, making it a good choice to overseed into fescue or brome pastures or hay meadows that, over time, have become more acidic. In southeastern Kansas, alsike clover has been used to patch-in newly established alfalfa fields in wet areas (such as terrace channels) where alfalfa refuses to grow.

Mixtures of alsike and red clover make good quality hay. Although alsike clover is lower yielding than red clover, its ability to withstand excessive soil moisture and its greater tolerance to acid soils helps ensure stand establishment. In many cases an alsike and red clover mixture will produce more hay combined than when planted separately.

Alsike clover can and will cause bloat issues and should be fed to livestock with care. On pastures high in alsike clover content, introduce animals gradually to the forage and never turn hungry cattle loose on alsike clover. It has been linked to "alsike poisoning" in horses and for that reason, alsike clover is never a good idea for horses.

Varieties

Only common or "variety not stated" alsike clover is commercially available in Kansas.

ARROWLEAF CLOVER

Arrowleaf clover is a cool season, reseeding annual legume. It is often utilized for grazing, hay production, a wildlife food source, soil improvement, and a cover crop. Forage quality is very good and antidotal information suggests that arrowleaf clover has a much lower potential to cause bloat than most other clovers. It can grow to a height of 40 to 50 inches under good conditions. In addition to cattle, deer and turkey seem to love arrowleaf clover.

The southern border of Kansas is considered the northern extant of arrowleaf clover's adaptation. However, excellent results have been obtained in southeast Kansas utilizing this productive legume.

Establishment

Best results are generally obtained by planting on prepared seedbeds in late August or early September. Planting rates range from 10 pounds per acre drilled to 15 pounds per acre broadcast. Producers in southeast Kansas have had success no-tilling arrowleaf clover into bermudagrass in September. Arrowleaf clover plants grow slowly in the fall and winter, but grow rapidly from February through May. This clover is capable of producing large seed yields with a high percentage of hard seed, which allows arrowleaf clover to maintain stands for several years.

Use and Management

Arrowleaf clover will continue to develop new leaves and remain productive long into the spring when grazed to a height of 2 to 4 inches. If managed for hay, arrowleaf clover should be grazed until early April, at which point livestock should be removed. Cut arrowleaf clover for hay at the bloom stage (usually mid-to-late May in Kansas).

Varieties

In Kansas, Yuchi is generally the commercial variety available. Arrowleaf clover may not survive winters in Kansas and should only be considered in the southernmost counties.

SWEET CLOVER

Sweet clover is native to Eurasia and was introduced to North America in the early 1700's. There are two principal types: yellow sweet clover (M. officinalis) and white sweet clover (M. alba). Generally, the cultivated forms of sweet clover are biennial; however, a few annuals such as Hubam, are also cultivated.

Sweet clover is adapted to all of eastern Kansas and to areas of western Kansas with favorable moisture conditions. Seedlings are weak and slow to establish, but once established, sweet clover is one of the most drought-resistant legumes.

In the seedling year, the biennial sweet clovers produce a plant with a single branched stem which obtains 12 to 36 inches of growth. The second year's growth may reach a height in excess of 6 feet. At the end of the second season, growth is completed, seed is produced, and the roots and tops die. From midsummer to early fall of the first year, buds are formed on root crowns. These remain dormant during winter and produce a strong, coarse growth of stems early the next season. Numerous small flowers are borne on racemes. Seed pods generally contain one seed.

Yellow sweet clover is considered to be better adapted to Kansas than is white. Yellow is more tolerant to adverse conditions such as drought and competition from companion crops. It also is slightly finer stemmed and generally superior to white in forage quality, none of which makes any difference to the beekeeper who prizes white sweet clover for the flavor it imparts to honey.

Establishment

In Kansas, sweet clover is usually broadcast into fall planted small grains or cool season pastures in February or March. Freezing and thawing effectively ensures adequate seed/soil contact for germination. Broadcast planting rates range from 12 to 20 pounds per acre. If planting date is delayed beyond what would be considered ideal, sweet clover should be drilled to ensure adequate soil coverage and to promote uniform, rapid emergence. When drilling, 10 to 14 pounds per acre is usually sufficient.

Sweet clover has a high percentage of seed with "hard" seed coats that are nearly impermeable. To insure good seedling emergence, seed should always be hulled and scarified. Use only good quality seed, (80% germination or higher). Inoculate with the proper strain of rhizobia bacteria and do not incorporate seed too deeply. Seed should be covered with no more than 1/4 inch of soil. If placed too deeply, poor emergence will result.

While sweet clover tolerates alkali conditions, it will not tolerate acid soils. Soil should be limed prior to planting sweet clover to a pH of 6.4 or higher.

Use and Management

Sweet clover is used for soil improvement, hay, silage, pasture and the production of honey. It is tolerant to poor drainage, overflow, and alkali conditions. It will grow on a wide range of soil types including heavy clays and rocky terrain.

As a soil improvement crop, sweet clover excels. Its massive tap root is capable of breaking up plow pans and penetrating and loosening soil to great depths. When plowed under, sweet clover can provide up to 100 pounds of nitrogen per acre to succeeding crops.

Sweet clover utilization in Kansas as pasture far exceeds its usage as hay or silage. As pasture, sweet clover produces forage high in protein but lacking in palatability. Coumarin, an aromatic compound, affects the palatability of sweet clover until livestock become accustomed to the bitter taste. If livestock are placed on sweet clover pasture when plants are young and small, acceptance is relatively good.

During heating and spoilage of sweet clover hay or silage, coumarin breaks down into a toxic substance (dicoumarel), which reduces the blood clotting ability of animals consuming the forage. Death may result.

New forage growth the second season does not come from the crowns as it does in alfalfa. Instead, growth originates from buds on the lower portion of the stem. Therefore, if cut too low, the second season regrowth will be prevented and the plants will die.

Varieties

The most popular sweet clover planted in Kansas is "common" of either local or Canadian origin. This common seedstock generally performs satisfactorily. If available, the variety Madrid yellow sweet clover is recommended for Kansas. There are cultivars that have been developed which are low in coumarin, such as Polara white sweet clover and Norgold yellow sweet clover. However, these varieties are not well-adapted as far south as Kansas.

KOREAN LESPEDEZA

Korean lespedeza was introduced to this country in 1919 from Korea. Day-length and temperature influence the vegetative growth, floral initiation, and seed maturation. It is a short-day cultivar. That is, plants remain vegetative until day length shortens sufficiently to initiate flowering. When Korean lespedeza is grown at more northern latitudes, floral initiation and seed maturation are delayed because its critical photoperiod occurs late in the season. If taken too far north, frost will kill plants before they mature seed. In the more southern latitudes, the high temperatures become the limiting factor in the use of Korean lespedeza. High temperatures reduce the number of days between flowering and maturity. Vegetative growth is retarded when seedlings emerge before the day lengths in spring are equal to or greater than the critical photoperiod. Thus, Korean lespedeza has a rather narrow band of adaptation, from northeast Oklahoma, through eastern Kansas, to southern Iowa and east to the Atlantic Ocean.

Unlike red clover or alfalfa, Korean lespedeza is a warm-season annual legume. Flowers appear in late summer to early fall and vary in color from light pink to purple. Leaves of Korean lespedeza turn forward around the developing seed, providing excellent protection against shattering.

When vegetative, it is fine-stemmed and leafy. Korean lespedeza possesses a shallow tap root system. Capable of producing nitrogen in excess of its own requirements, Korean lespedeza is an excellent supplement to cool-season pastures in eastern Kansas.

In spite of the fact that it is an annual, Korean lespedeza is very persistent. Natural reseeding from hard seed results in good stands year after year as long as competition is not too severe.

Establishment

Korean lespedeza may be sown from midwinter to early spring. Winter broadcasting without covering normally results in adequate stands in pastures, on idle ground, and in small grains. Unhulled Korean lespedeza should broadcast at 8 to 15 pounds per acre in pastures and 15 to 25 pounds per acre in small grains. Use the lower end of the seeding rate if sowing as a conservation practice or if Korean lespedeza has existed in the same field the previous year and some volunteer is expected. Use the upper end of the seeding rate to maximize hay or seed yields. Seed should be inoculated with the proper strain of rhizobia bacteria prior to planting.

Korean lespedeza seed will not germinate well shortly after harvest. Seed with "hard" seed coats will range from 40% to 60% during the first two months following harvest in October, but will decline to 15% or less by February.

Use and Management

A substantial portion of land used for grazing in the upper southeastern and mid-eastern United States contains Korean lespedeza, either in pure stands or in mixes. Korean lespedeza performs especially well on infertile, marginal soils, where little or no nitrogen fertilizer is applied to perennial grasses. In its area of adaptation, Korean lespedeza is widely used for pasture, hay, soil conservation and wildlife plantings.

Korean lespedeza is highly regarded for its high nutritional qualities and good animal performance. Hay yield of Korean lespedeza generally range from 1 1/2 to 2 1/2 tons per acre. It should be cut in the early-bloom stage for best quality. Cut timely and properly cured, Korean lespedeza hay is nearly equal to alfalfa hay in feed value.

When utilized for pasture, Korean lespedeza is most often interseeded into cool-season perennial grasses. This association can be very successful. However, the grasses must be managed to reduce competition with the legume. Heavy nitrogen fertilization of grass sods can be extremely detrimental to the persistence of Korean lespedeza. When grown in association with a cool-season legume such as alfalfa or ladino clover, the combined (warm and cool-season) legumes can often supply an adequate quantity of nitrogen for the grass.

Korean lespedeza can make a tremendous contribution on pastureland at relatively low levels of maintenance or on soils of relatively low fertility. It is especially valuable in association with cool-season grasses because it provides high quality forage in summer and early fall. It's also valuable in revegetating surface mine spoils, where it reseeds and spreads.

Varieties

In Kansas, Korean lespedeza seed of local origin is preferred. Not only is the producer assured of well adapted seed, Kansas seed laws provide truth-in-labeling statutes that protect producers from excessive and troublesome weed seeds. Unfortunately, farmers in other states which produce Korean lespedeza seed are not afforded the same level of consumer protection.

Kobe lespedeza is another annual lespedeza that is sometimes utilized in Kansas. However, Kobe is later maturing than Korean and may not reliably produce a seed crop before the first killing frost of autumn.

LADINO CLOVER

Ladino clover is a large or giant type of white clover. There are also small and intermediate types of white clover. The small type is usually referred to as "Wild White." The small type produces low yields, and is seldom planted in improved pastures. The intermediate type of white clover has commonly been referred to as "White Dutch Clover." It is intermediate in size to the small type and ladino. White Dutch clover generally flowers earlier and more profusely than ladino and is preferred in those areas of the United States where white clover behaves as a reseeding annual. In Kansas, the large type (ladino) is much preferred over either the small or intermediate types.

Ladino clover excels under conditions of optimum soil moisture and fertility. It is shallow-rooted and spreads by solid stolons that root at the nodes. Leaves are trifoliate and usually are marked with a white *V*. Flowers are usually white, occasionally pinkish, and are borne on heads containing 20 to 150 individual perfect flowers. Plants flower and produce viable seed the first summer. Under stressful conditions, the persistence of ladino in a pasture depends upon volunteer seedlings. Under more favorable conditions, ladino persists as a perennial returning from the roots. In Kansas, persistence of ladino clover is a function of both.

Ladino clover is entirely cross-pollinated, relying upon honeybees and other pollen collecting insects. Pods usually contain 3 to 4 seeds, which mature three to four weeks after pollination.

Leaves and flowers are the only plant parts harvested. Consequently, the forage yield of the clover component of a ladino/grass pasture is relatively low. However, total forage yield, forage quality and distribution of production are excellent. Total yield of a ladino clover/grass pasture is comparable to that of other legume/grass mixtures.

White clover has been described as a "tonic" because of its beneficial effects on livestock grazing it. Ladino is low in fiber and is highly digestible, nutritious, and palatable. On a dry weight basis, crude protein can run 30% and digestibility may range from 60% to 80%. Ladino clover has been credited with improved animal health, increased milk flow, increased calf weaning weight and daily gains, and improved conception rates.

Establishment

Often, the major problem in managing a ladino/grass pasture is establishment and maintenance of the clover in the desired proportion. Generally, the legume should contribute from 20% to 40% of the total forage yield of the mixture. Obtaining a good stand of clover initially is half the battle. It is best to establish the legume at the time of grass establishment. However, ladino clover may be established in existing sods, although less reliably so than on a well-prepared seedbed.

When establishing the grass and legume simultaneously, the seedbed needs to be fine on top, firm and free of weeds and plant residue. Planting rates of 1.5 to 3 pounds per acre of good quality inoculated ladino seed is adequate. Ladino clover seed is small, approximately 800,000 seeds per pound, so higher seeding rates are not necessary. Ideally, the grass should be drilled with every other drop tube plugged and the entire area overseeded with clover seed.

Soil pH should be higher than 6.3 and phosphate and potassium levels in the soil should be in the high range.

In Kansas, ladino clover may be seeded either in late summer/early fall or late winter/early spring. Ladino clover must have a strong root system before it is exposed to stresses from winter freezing or summer heat and drought.

When establishing ladino clover in existing sods, fertility requirements must be met and the sod lightly disked or grubbed down where clover seed can make good soil contact. The grass should not be fertilized with nitrogen during the season of clover establishment. Most grasses are more competitive for potassium than ladino clover. Even when normal amounts of potash are applied, the grass may cause potassium deficiency in the clover. When nitrogen levels in the soil are low and other minerals are adequate, ladino will compete with most grasses.

A frequent mistake in the establishment of ladino clover is to allow the grass to shade the clover. Clipping in early spring will help control grass and weeds and maintain a favorable environment for the clover.

Use and Management

Ladino clover is well suited to the eastern fourth of Kansas. In tall fescue, ladino is generally more persistent than other perennial legumes. This is due to its capacity to reseed itself and not rely solely upon its perennial growth habit.

Ladino clover in cool-season hay meadows will increase total forage yield and quality. Management of the grass/legume mixture must be centered on the maintenance of the legume.

Use of ladino clover for pasture is much more widespread than its usage for hay. Ladino clover is well suited to rotational grazing or continuous grazing. Rotational grazing provides maximum animal performance but requires added management expense. The "rule-of-thumb" is to allow forage growth to reach a height of 8 to 12 inches and then graze to 3 to 4 inches.

Controlled continuous grazing is the practice most often utilized in Kansas. Grazing pressure should be controlled by either stocking rates or varying the length of time animals are allowed to graze per day to maintain forage height between 4 and 8 inches. Ladino clover/grass pastures must be managed so that the carbohydrate reserves in the clover roots are not depleted prior to the winter season.

Varieties

One of the most highly regarded varieties of ladino clover for Kansas is Regal. Most ladino clover seed planted in Kansas is "common" from California. This common seedstock performs satisfactory at a lower cost to the customer. However, only quality seed from reliable sources should be purchased.

BIRDSFOOT TREFOIL

Birdsfoot trefoil is a long-lived perennial legume that grows on many varied soil types (from clay to sandy loams). It will grow on shallow, infertile, acid, or mildly alkaline soils, mined land, and tolerate water-logged conditions. However, it is most productive on fertile, moderately well drained soils having a pH of 6.4 or higher.

In Kansas, birdsfoot trefoil is limited to the eastern 1/4 of the state. Only the "Empire" type cultivars should be used.

Mature plants of birdsfoot trefoil have many well-branched stems arising from a single crown. Under favorable conditions, the main stems may obtain a length of 24 to 36 inches. The leaves are compound and alternately attached on opposite sides of the stem. During darkness, the leaflets fold around the petiole and stem. When cut and dried for hay, the leaves wrap around the stems, giving the impression of excessive leaf loss.

The inflorescence is an umbel (like a carrot), having four to eight florets. Flower color ranges from light to dark yellow and may be tinged with orange and red stripes. Pollination of flowers is dependent upon insects. Ten to fifteen seeds are borne in long cylindrical pods, which turn brown to almost black at maturity. Three to five pods are attached at right angles to the end of the flower stem and are arranged so that they impart the appearance of a bird's foot. Pods ripen 25 to 30 days after pollination. They split along both sides when mature and twist spirally to scatter the seed.

Birdsfoot trefoil possesses a well-developed tap root system that penetrates deeply. There are numerous lateral branch roots in the upper 15 inches of soil. Roots have the ability to produce new shoots and roots.

The nutritive value of birdsfoot trefoil is equal to or exceeds that of alfalfa. Hay yields are usually 50% to 80% that of alfalfa.

Establishment

Seed should be mixed with ample amounts of the proper rhizobia bacteria inoculant immediately before planting. The growth rate of birdsfoot trefoil seedlings is relatively slow. Because of the small seed size and slow seedling growth rate, seedbeds must be prepared carefully. The seedbed surface must be smooth and firm. Seed should be incorporated no deeper than 1/4 inch. Packing the soil before and after sowing improves contact of the seed with soil moisture and can result in quicker, more uniform emergence.

Both spring and late summer seeding dates can be successful. In Kansas, late summer seeding is preferred. Late summer plantings generally do not suffer as much competition from weeds as do spring plantings, and a late summer seeding date allows birdsfoot trefoil to become fully established the following spring before the onset of another hot, dry period. Both broadcast and plantings with a Brillion drill or similar equipment have been successful.

Recommended seeding rates range from 5 to 10 pounds per acre. The higher seeding rates often promote establishment of more seedlings and greater forage yield the first year but, have little effect on forage yields in succeeding years.

Birdsfoot trefoil is routinely used to renovate grass sods. Animal production (gain) usually is increased and economic return can be greater than that from grass fertilized with nitrogen. Good stands have been obtained in southeastern Kansas by broadcasting a mixture of seed, phosphate, and potash fertilizer. The grass sod should be lightly tilled or grubbed down and should not be fertilized with nitrogen the year of establishment.

Use and Management

In Kansas, usage of birdsfoot trefoil for pasture far exceeds its utilization as hay or silage. Once established and properly managed, it is very persistent and can increase pasture yield four-to-fivefold. Birdsfoot trefoil blooms intermittently throughout the summer. It often sets seed on low-growing stems, even when closely grazed. This ability to reseed itself greatly enhances its ability to persist year after year. Birdsfoot trefoil pasture can be stockpiled because it retains its high quality leaves on mature growth. Also, birdsfoot trefoil continually produces new shoots from axillary buds as the stems mature.

Rotational rather than continuous grazing is preferred. Rotational grazing systems allow long periods of photosynthesis to build root carbohydrate reserves between defoliation periods. Since carbohydrates reserves stored in the roots are relatively low under warm to hot growing conditions, the rest periods provided by rotational grazing are very important in Kansas.

When utilized for hay, birdsfoot trefoil can provide two or three crops per year, depending upon growing conditions. While pure stands provide excellent quality forage, associated grasses contribute to higher forage yields by filling in vacant areas. Grasses in association with birdsfoot trefoil also reduce lodging, decrease severity of soil heaving, and speed hay harvesting and curing. Non-competitive grasses that are of high quality, such as timothy, do not depress the forage yields of birdsfoot trefoil, and generally result in greatest success.

Varieties

In Kansas, the variety most commonly grown is Empire, which was the first birdsfoot trefoil cultivar developed and one that still performs satisfactorily.

CROWN VETCH

Crown vetch is a long-lived perennial legume which spreads by creeping underground rootstocks. It possesses a deep penetrating taproot and numerous branch roots. Stems are hollow, semi-decumbent and may reach a length of 4 feet. Flowers range in color from white to purple with various shades in between. Seeds are borne in pods that are divided into three to twelve single seeded sections. Seeds are mahogany in color and are rod-shaped.

Crown vetch is widely renowned for its ability to hold soil on severe slopes and is routinely used on highway right-of-ways. Valued for erosion control on steep embankments, mine spoil areas and other disturbed sites; crown vetch is potentially an important forage species for eastern Kansas. It also serves as an ornamental ground cover.

While established plants are tolerant of moderately acid and infertile soils, it is best suited to well-drained, fertile soils with a pH of 6.4 or higher.

Establishment

Seed of crown vetch have a hard seed coat and must be scarified before planting. Inoculate seed with the proper strain of rhizobia bacteria prior to planting. Seedling vigor of crown vetch is poor, but satisfactory stands can be established by sowing into a firm seedbed. Competition from weeds and grasses grown in association with crown vetch must be minimized.

While difficult, it is not impossible to establish crown vetch in established grass sods. It is critical to manage the grass to favor the growth and development of the legume.

Broadcast seeding in late summer or early spring of 4 to 8 pounds seed per acre followed by harrowing or cultipacking is generally recommended. Have patience – crown vetch is a slow starter but spreads via underground rootstocks as well as seed. It may take a couple of years to achieve desired stands.

Use and Management

In Kansas, crown vetch is most commonly utilized for pasture. Forage yields are equal to or superior to other more commonly grown pasture legumes. Improper grazing management can greatly reduce stand persistence. Rotational grazing, with rest periods sufficiently long enough (30 to 40 days) to replenish carbohydrate reserves in the roots, are necessary to maintain stand productivity.

When utilized for hay, two harvests per season may be attempted. Seasonal trends of carbohydrate reserves in the roots are similar to those of alfalfa. The first hay crop should be taken when crown vetch is in full bloom. This allows better regrowth as the number of axillary buds producing regrowth is highest at this time.

In addition to crown vetch's utilization as a forage legume, it is also widely used for erosion control, highway beautification, ornamental ground cover and wildlife plantings.

Varieties

In Kansas, the most widely recommended variety is Emerald. It was developed in Iowa and has proven to be well adapted to the soils and climate of eastern Kansas.

HAIRY VETCH

Hairy vetch is cold-hardy annual legume adapted over a wide area of the United States. Hairy vetch tolerates poorly drained and acid soils much better than most other legumes. Plants are vining with stems that attain lengths of 2 to 6 feet. The stems bear pinnate leaflets and terminate in tendrils that wrap tightly around surrounding plants. Purple flowers are borne in clusters and pods are elongated and compressed. Seed is round or oval and grayish black in color. The name hairy vetch is a misnomer as plants are nearly hairless.

Hairy vetch seeds shatter quickly after pods mature. However, it is fairly "soft-seeded" (having little or no seed dormancy) and does not dependably reseed; therefore, it should be replanted every year.

Hairy vetch can increase carrying capacity and quality of grass forage substantially, as well as reduce the need for commercial nitrogen fertilizer. With nitrogen fertilizer prices expected to escalate in price in coming years, hairy vetch usage will likely increase.

Establishment

Hairy vetch seed should be inoculated with the proper strain of rhizobia bacteria prior to planting. If seeding with a winter small-grain, 10 to 20 pounds per acre is adequate. If broadcasting into permanent pasture 30 to 40 pounds per acre is recommended and if drilling into grass sods 20 to 25 pounds per acre is sufficient. Only good quality seed (80% or higher germination) should be used and should not be placed deeper than 1 inch.

Little germination of seeds occurs at temperatures above 70° F, but good germination occurs between 40° F and 70° F. In Kansas, hairy vetch should be planted as early as possible in the fall to promote early growth for winter pasture.

While hairy vetch is tolerant of acid soil, extremely acid subsoils will severely restrict root growth. It has a high phosphorus requirement and any deficiency of this mineral should be corrected prior to planting.

Use and Management

Hairy vetch is widely utilized for pasture and hay. In Kansas, hairy vetch is often seeded with winter rye or winter wheat for winter pasture. The following spring this mixture is either grazed out, or cut for hay, haylage or silage. Rye generally produces more forage in the fall than wheat; however, wheat can provide up to two weeks longer grazing in the spring. Hairy vetch used in this manner serves to increase both forage yield and quality.

Likewise, hairy vetch can be seeded into permanent pasture to increase carrying capacity and forage quality. In pasture situations, grazing of hairy vetch should not begin until the plants are at least 6 inches tall. It should not be grazed lower than the lowest leaf axil, as axillary buds will be removed, greatly slowing regrowth. If hairy vetch is utilized for hay, the optimum yield and quality is obtained when cut in the early-bloom stage.

Recently, much attention has been given to using hairy vetch as a biological nitrogen producer for row crops. In this system, hairy vetch is seeded in pure stands and is grazed through the winter and early spring. It is then treated with an appropriate contact herbicide and corn or grain sorghum is no-tilled into the vetch. Hairy vetch utilized in this manner can produce 80 to 100 pounds of nitrogen per acre for subsequent crops.

Varieties

Hairy vetch seeded in Kansas is almost exclusively "common" or "variety not stated." Nebraska University has developed a high-yielding variety, Madison, which should work well in Kansas. Most hairy vetch seed production takes place in Oregon. However, Oklahoma and surrounding states often produce enough seed to meet local demands.

WINTER FIELD PEAS

The winter field pea is a cool-season annual legume with sufficient winter hardiness for the southern half of Kansas. Similar in appearance to the garden pea, winter field peas are vining with seeds borne in pods. Flowers are purple and seed color is mottled gray. Root growth is massive providing excellent erosion control and soil improvement.

Well-drained clay loam soils of limestone origin are best for field pea production. Soil drainage should be good to excellent as field peas will not tolerate water-logged conditions. Therefore, heavy clay soils should be avoided. High temperatures can severely injure the crop. Field peas perform best in cool temperatures when rainfall is fairly abundant during the fall, winter and spring months.

A very versatile crop, winter field peas are utilized for hay, silage, crop rotation, cover crop and soil improvement. Planted in late summer (August/September), the crop protects the soil from erosion during the winter and spring months and then can be turned under as a green manure crop.

Establishment

Winter field peas do not tolerate acid soils; pH values should be 6.2 or higher for successful production. Likewise, they have a high requirement for phosphate. Any deficiencies in these areas should be corrected prior to attempting to establish field peas.

In Kansas, field peas should be planted in late August or September. These planting dates give seedlings time to develop a strong root system before the onset of winter, reducing the danger of winter injury and promoting more rapid growth in the spring.

Winter field peas should be drilled at a rate of 35 to 45 pounds per acre into a well-prepared firm seedbed. Field peas will emerge from relatively deep plantings. However, a 1 to 2 inch seeding depth is best as it allows more rapid growth and uniform emergence. Seed should always be inoculated with the proper strain of rhizobia bacteria prior to planting. Inoculated seed produces plant roots that are quickly colonized by bacteria, enabling the plant to fix atmospheric nitrogen. As a result, plants are healthy and vigorous, greatly enhancing their ability to improve the soil.

Use and Management

In Kansas, winter field peas are most often used as a cover crop, protecting the soil from erosion during the fall and winter. The following spring (late March or April), the peas are turned under as a green manure crop. In this management system, much greater benefit will be achieved if plants are allowed to reach maximum growth. Forage yields often double by delaying harvest from early March until April. Winter field peas managed in this manner will provide roughly 40 pounds nitrogen for subsequent crops.

Winter field peas produce a forage of very high quality. When peas are utilized as a forage legume, best performance is obtained by growing them in association with small grains. Winter pea/small-grain combinations provide excellent grazing, hay, or silage. The grain stems help support the pea vines, reduce lodging and make a better-balanced feed. For optimum hay or silage production and quality, harvest should begin when the first pods begin to appear.

Varieties

Austrian Winter, a common cultivar is most commonly grown in Kansas. Dixie Wonder is a variety selected out of Austrian Winter that is well-suited to southeastern Kansas. Dixie Wonder is seven to ten days earlier in maturity but forage yield is usually less.

CHAPTER SIX: BRASSICAS

Brassicas are cold-hardy, high quality, fast growing crops that are suitable for livestock grazing. There are many species of brassica, such as 1) Kale: Kale is the most winter-hardy of all the brassicas. It is not a root crop, but rather is grown for its highly palatable leaves and stems. It generally takes 150 to 180 days to reach maximum production and average protein content ranges from 15% and 17%. 2) Rape: Rape is a multi-stemmed forage with a fibrous root system. It is generally ready to graze 60 to 90 days after emergence. Plants are ready for harvest when leaves change to a purplish or bronze-tinge color. There are two types of rape: a giant type which is leafy and upright and a dwarf type which is short and branched. The giant type is best suited to cattle grazing due to its higher palatability. The dwarf type is most often utilized for finishing lambs or for wildlife plantings. 3) Swede: Swede is a very long-season plant with an edible root. It generally requires 150 to 180 days to reach peak production and is best suited to late fall grazing. 4) Brassica Hybrids: Hybrids are crosses between various brassica species. Most hybrids are crosses involving turnips and various oriental vegetables. Most are very leafy and can be grazed earlier than other brassicas. However, they vary greatly in their ability to maintain palatability and quality. 5) Turnips: Turnips are a relatively short-season brassica that provides roots (bulbs), stems, and leaf growth for rotational or strip grazing. Turnips have bushy tops and large white roots that are rich in carbohydrates. In Kansas, turnips are the most economically viable brassica forage crop.

Brassicas are very high in crude protein and energy and low in fiber. Turnips, for example, have leaves that normally contain 20% to 25% crude protein, 65% to 80% in vitro digestible dry matter (IVDDM), about 20% neutral detergent fiber (NDF), and about 23% acid detergent fiber (ADF). The roots contain 10% to 14% crude protein and 80% to 85% IVDDM.

Turnips and other brassicas can provide grazing at any time during the summer and fall depending upon planting date. In Kansas, turnips are most often planted in July or August for late fall grazing.

Turnips will maintain their forage quality, if not headed, even after freezing temperatures. Rape is most often sown in the spring and utilized for summer pasture. Turnips generally require 75 to 90 days to reach their maximum potential forage yield, while rape forage yield usually peaks in about 120 days after emergence.

The high digestibility of brassicas allows increased rate gains over many other forage crops. Brassicas are highly palatable and can significantly extend the effective grazing season. Turnips will continue to grow until temperatures drop as low as 15^{0} F. Several days of below freezing temperatures are required to kill turnips.

The brassicas can be easily grown with little or no cultivation and are frequently used to support beef, dairy, sheep, and even swine production.

TURNIPS

Turnips generally have large global or tapered roots and are rich in carbohydrates. Forage turnips have been selected that set roots so that portions are exposed above the soil surface, allowing grazing livestock access to both top growth and roots. Foliage is erect and succulent during vegetative growth. Turnips generally require 10 to 12 weeks of growth prior to grazing.

Turnips require soils with good drainage and a pH range of 5.3 to 6.8. They may be planted in spring or late summer. In Kansas, turnips are generally planted in August and utilized as a high quality forage for beef cattle in the fall and early winter. Planting dates in September are often successful, but forage yield will be reduced.

Establishment

Turnips have approximately 240,000 seeds per pound. This small seed size makes them economical to plant, but also makes it difficult to calibrate drills or broadcast seeders. 1 1/2 to 3 pounds per acre is sufficient to achieve desired stands when drilled. If broadcasting seed, it is best to increase seeding rate to 2 to 4 pounds per acre. Turnips must not be planted too deeply. Generally 1/4 inch is ideal, and never exceed 1/2 inch depth.

Turnips sown in May or June will provide a quality forage for grazing livestock in August and September. Planted in July or August, turnips may be utilized in November and December and have been known to hold forage quality well into January.

Use and Management

Turnips do best on good soil that drains well; they do not tolerate water-logged soil conditions.

Fertility requirements for turnips are similar to wheat; however, excessive nitrogen can result in dangerous levels of nitrate. Generally, 75 to 90 pounds of actual nitrogen is sufficient to promote high forage yield and high crude protein. Phosphorous is often a limiting factor in turnip production. All brassica crops require a good deal of phosphorous to be productive. Generally, 40 to 45 pounds per acre is recommended, depending upon soil type and soil phosphorous levels. Potassium is seldom a limiting factor in Kansas. However, the only way to determine a deficiency is to soil test.

Hungry animals should never be turned out on lush turnip pasture. In fact, producers should either put cattle on a protein and energy rich diet two to three weeks prior to grazing turnips, or introduce livestock to turnips gradually. For the first week to ten days, allow cattle to graze turnips for a limited time. At the beginning, allow two to three hours per day and gradually increase the allotted time until cattle become fully accustomed to turnips. Animals may take a few days to acclimate, but once accustomed to turnips, gains of 1 1/2 to 2 pounds per day for 500-pound calves or 1/4 to 1/3 pound per day for lambs are not unusual.

For maximum production and animal performance, turnips should be strip grazed. Cordon off an area with electric fencing and once this area is fully depleted of leaves and roots, move the fence and livestock to an adjacent area. Grazing in this manner will minimize losses due to trampling and is the most efficient use of available forage.

A good, clean water supply is necessary to ensure the animals' appetite is not suppressed and ensure the metabolic requirements are met. It is also a good idea to feed dry roughage with turnips. Two or 3 pounds of hay or straw per animal should be fed daily. An alternative would be to allow animals' free access to a dry pasture or corn stalks adjacent to the turnip pasture. Copper, manganese, and zinc contents of turnips do not meet dietary requirements of livestock, so mineral supplements are required. Also, iodine, iron and copper supplements will help prevent anemia and goiter.

In southeast Kansas, there has been a good deal of interest in using turnips to supplement tall fescue pastures. While it is difficult-to-nearly-impossible to achieve desirable stands of turnips in lush, vigorous tall fescue pastures, sowing a couple of pounds of turnip seed into overgrazed or drought-stricken tall fescue is often successful. Turnips will extend the grazing season, increase carrying capacity, and improve animal performance.

Turnips sowed in mixture with small grains makes for excellent fall and winter pasture. When sowing small grain/turnip mixtures, producers should broadcast the turnip seed prior to drilling the wheat, oat, rye, or barley component. Do not mix the seeds and drill together, as this will result in the turnips being planted too deeply to emerge. Running the drill over the top of broadcast turnip seed will help ensure good soil/seed contact and help ensure a uniform stand. It is also wise to drill the small grain at about 1/2 the normal seeding rate to reduce competition with turnip seedlings.

Perhaps the most effective utilization of turnips in Kansas is a double-crop, sown into corn stalks or wheat stubble. While there are many management factors to consider, turnips are an excellent forage at a time when most grasses are unproductive or unavailable. Most of the issues with turnip forage can be overcome by introducing grazing animals gradually, not allowing hungry animals to gorge on turnip forage, feeding a couple of pounds of dry hay per animal per day, and providing supplemental minerals (especially iodized salt).

Varieties

There are many varieties of brassica that are promoted for use in Kansas. Our experience has been that the old Purple Top White Globe variety is tough to beat.

DAIKON RADISH

Another member of the brassica family that has received much attention in recent years is Daikon radish. Like all the forage brassicas, Daikon radish produces forage high in protein with good digestibility. However, cattle are selective grazers and, if alternative forage is available, they will almost always reject the radish plants in favor of the alternative – regardless of quality. Likewise, grazing Daikon radish may diminish or impair this remarkable plant's ability to provide unique and valuable soil improvement attributes.

Daikon radish cover crops are often used to reduce the effects of soil compaction. These plants have amazing taproots. The thin lower portion of the taproot can extend into the soil 6 feet or more in the fall. The thick, fleshy portion of the upper taproot may grow 12 to 20 inches long with a diameter of 2 to 5 inches or more. These thick, fleshy upper roots often protrude 2 to 6 inches above the ground and create vertical holes that effectively break-up surface soil compaction and greatly improve soil tilth. The radish plants die in the winter and the roots decompose, leaving open root channels that a spring planted row crop may utilize, allowing that crop's roots to follow those channels and grow through compacted soil layers.

The taproots of Daikon radish also mine nitrogen and other nutrients from deep within the soil profile and pull them up to near the soil surface. These nutrients are stored in the fleshy upper roots and above ground plant tissues. As the radish plants die and plants decompose in January and February, these nutrients are released back into the root zone where they are available for a spring-planted crop. This can significantly reduce the need for expensive commercial fertilizer (studies have documented Daikon radish's ability to capture and release over 100 pounds of nitrogen). It also effectively prevents excess nitrate from leaching into groundwater and streams.

Daikon radish plants grow rapidly when planted in late summer or early fall and can provide full canopy closure within three to four weeks. This dense canopy can eliminate weed emergence in the fall and winter and often results in a weed-free seedbed the following spring. Radish residue is sparse and fragile, making it possible to no-till a spring planted crop without a burn-down herbicide application. Because the residue is slight and the soil surface is punctured by large root holes, the seedbed warms and dries considerably faster in the spring. The warmer, drier soil and the elimination of the need for tillage fosters earlier spring planting.

Generally, 8 to 10 pounds per acre is recommended if drilling and 12 to 14 pounds per acre if broadcasting. If mixed with turnips or another cover crop seed, planting rates may be reduced by as much as 50%. Planting at lower rates will result in the fleshy portion of the root becoming larger in diameter. In Kansas, Daikon radish should be planted from August through early September.

While Daikon radish may be utilized for forage, its true value is as a cover crop. It is not unusual for total dry matter production (shoots and roots) to exceed 3 tons per acre. This crop will add significant quantities of easily decomposed organic matter to the soil and through its beneficial effect on soil structure, tilth, and fertility; it can provide significant yield improvement for spring planted row crops.

There are many brands of Daikon radish being offered in Kansas. They all are essentially the same.

QUICK SEEDING CHART

Species	Seeds/lb	lbs/Acre	Planting Date
NATIVE GRASSES			
Little Bluestem	238,000	4 to 6 PLS lbs	Nov 15th – May 15th
Big Bluestem	165,000	5 to 8 PLS lbs	Nov 15th – May 15th
Sand Bluestem	120,000	5 to 8 PLS lbs	Nov 15th – May 15th
Switchgrass (upland)	250,000	3 to 5 PLS lbs	Nov 15th – May 15th
Switchgrass (bottom)	500,000	2 to 4 PLS lbs	Nov 15th – May 15th
Indiangrass	133,000	5 to 8 PLS lbs	Nov 15th – May 15th
Sideoats Grama	160,000	4 to 6 PLS lbs	Nov 15th – May 15th
Blue Grama	725,000	2 to 4 PLS lbs	Nov 15th – May 15th
Buffalograss	50,000	6 to 10 PLS lbs	Nov 15th – May 15th
Eastern Gamagrass	7,500	8 to 12 PLS lbs	Nov 15th – May 15th
Western Wheatgrass	115,000	6 to 10 PLS lbs	Aug 20th – May15th
INTRODUCED GRASSES			
Smooth Bromegrass	140,000	15 to 25 lbs	Aug 20th – Oct 1st Mar 15th - Apr 30th
Tall Fescue	200,000	20 to 30 lbs	Aug 20th – Oct 1st Mar 15th - Apr 30th
Orchardgrass	420,000	10 to 20 lbs	Aug 20th – Oct 1st Mar 15th - Apr 30th
Annual Ryegrass	190,000	20 to 30 lbs	Aug 20th – Oct 1st Mar 15th - Apr 30th
Perennial Ryegrass	240,000	20 to 30 lbs	Aug 20th – Oct 1st Mar 15th - Apr 30th

Species	Seeds/lb	lbs/Acre	Planting Date
Timothy	1,200,000	4 to 8 lbs	Aug 20th – Oct 1st Mar 15th - Apr 30th
Reed Canarygrass	530,000	10 to 15 lbs	Aug 20th – Oct 1st Mar 15th - Apr 30th
Kentucky Bluegrass	1,400,000	4 to 6 lbs	Aug 20th – Oct 1st Mar 15th - Apr 30th
Bermudagrass	1,000,000	15 to 20 bu (sprigs) 5 to 8 lbs (seed)	Feb 1st - June 1st
Crabgrass	825,000	3 to 8 lbs	Feb 1st - June 1st
Inter. Wheatgrass	75,000	10 to 15 lbs	Aug 20th – Oct 1st Mar 15th – May 30th
Pub. Wheatgrass	80,000	10 to 15 lbs	Aug 20th - Oct 1st Mar 15th – May 30th
Tall Wheatgrass	73,000	10 to 15 lbs	Aug 20th – Oct 1st Mar 15th – May 30th

WARM SEASON ANNUALS

Species	Seeds/lb	lbs/Acre	Planting Date
Teff Grass	1,400,000	8 to 10 lbs	May 20th – Aug 1st
Sudangrass	45,000	20 to 25 lbs	May 20th – Aug 1st
Sorghum/Sudan	18,000	25 to 40 lbs	May 20th – Aug 1st
Forage Sorghum	17,000	5 to 10 lbs	May 20th – Aug 1st
Hybrid Pearl Millet	90,000	10 to 20 lbs	May 20th – Aug 1st
Foxtail Millet	210,000	15 to 25 lbs	May 20th – Aug 1st

CEREAL SMALL GRAINS

Species	Seeds/lb	lbs/Acre	Planting Date
Winter Wheat	14,000	60 to 120 lbs	Aug 20th – Nov 15th
Oats	19,000	75 to 90 lbs	Feb 15th – Apr 10th
Barley	14,000	80 to 110 lbs	Aug 20th – Oct 31st
Winter Rye	18,000	60 to 100 lbs	Aug 20th – Oct 31st
Triticale	12,000	75 to 100 lbs	Aug 20th – Oct 31st

Species	Seeds/lb	lbs/Acre	Planting Date
LEGUMES			
Alfalfa	220,000	15 to 20 lbs	Aug 20th – Sept 30th Mar 15th – May 15th
Red Clover	270,000	8 to 12 lbs	Aug 20th – Sept 30th Feb 1st – May 15th
Alsike Clover	680,000	3 to 4 lbs	Aug 20th – Sept 30th Feb 1st – May 15th
Crimson Clover	150,000	10 to 15 lbs	Aug 15th – Sept 30th
Arrowleaf Clover	375,000	8 to 12 lbs	Aug 15th – Sept 30th
Sweet Clover	260,000	15 to 20 lbs	Feb 1st – Apr 30th
Ladino Clover	790,000	2 to 4 lbs	Aug 20th – Sept 30th Feb 1st – May 15th
Korean Lespedeza	240,000	10 to 30 lbs	Feb 1st – Apr 30th
Birdsfoot Trefoil	370,000	4 to 8 lbs	Aug 1st – Sept 20th Feb 1st – Apr 30th
Crown Vetch	110,000	5 to 10 lbs	Aug 15th – Sept 30th Feb 1st – Apr 30th
Hairy Vetch	16,000	15 to 20 lbs	Aug 15th – Oct 15th Feb 1st – Apr 15th
Winter Field Peas	2,500	20 to 30 lbs	Aug 15th – Sept 30th
BRASSICAS			
Turnip	167,000	3 to 6 lbs	Aug 15th – Sept 30th April 1st – June 1st
Forage Rape	160,000	3 to 6 lbs	Aug 15th – Sept 30th Apr 1st – June 1st
Daikon Radish	40,000	8 to 10 lbs	Aug 1st – Sept 15th

GLOSSARY OF RELEVENT TERMINOLOGY

Adventitious: Sprouting or growing from unusual places, such as roots originating from a stem.

Aerial: Plant structures originating above ground.

Alluvium: Sands, silts, etc. deposited by water.

Angiosperm: Flowering plant producing seed enclosed in a structure derived from the ovary.

Anther: The pollen bearing portion of the stamen.

Awn: A stiff bristle – in grasses usually situated at the tip of a glume.

Basal: Usually pertains to the base of the plant or plant structure.

Biennial: A plant that requires 2 years to complete its life cycle.

Capillary: Hair-like

Carpel: A pistil, or one of the units of a compound pistil.

Caryopsis: In grasses – a seed-like fruit.

Chlorophyll: The green photosynthetic pigment.

Cilia: Hairs or slender bristles.

Convolute: Rolled-up longitudinally.

Crown: That portion of a stem at the ground surface.

Culm: The stem of grasses, sedges, and rushes.

Cultivar: A cultivated variation (a variety).

Deciduous: Plants that shed leaves before a dry season or winter to minimize water loss. (not evergreen)

Decumbent: Trailing along the ground but with flowers or summit of stems erect.

Determinate: Inflorescence whose terminal flowers open first. Once flowering begins, plants stop growing.

Digestion: The conversion of insoluble, complex substances into soluble, simpler substances under the control of specific enzymes.

Dioecious: Pertaining to plants which have either staminate (male) or pistillate (female) flowers but not both.

Dormancy: The temporary cessation of growth under harsh conditions.

Ecosystem: The interaction of living organisms with each other and with the non-living environment.

Enzyme: A complex protein that promotes or signals a chemical reaction in living cells.

Epicotyl: That portion of a seedling above the cotyledon's attachment point.

Epidermis: A single layer of cells that forms the outer tissue of leaves, roots, and young stems.

Fiber: Cells which are long and thick walled, often containing protoplasm that is dead at maturity.

Fibrous Roots: Clusters of similarly sized roots, found in some dicots and most monocots.

Floret: Tiny flowers belonging to the inflorescence of the grass family.

Habitat: The natural environment in which plants complete their life cycles.

Hypocotyl: The area between the cotyledon and the radical in seedling plants.

Indeterminate: Plants whose terminal flowers open last. Plants continue to grow after flowering & flowering occurs over long a timeframe.

Internode: The region between two nodes.

Lateral Roots: Scores of small roots stemming from the tap root.

Meristem: Region of undifferentiated, actively dividing cells from which new cells emerge. (Growing points)

Midrib: The main middle vein of a leaf or leaflet.

Monoecious: Plants which possess both male and female flowers on the same plant.

Natural Selection: The process of evolution involving the population rise of organisms which have inherited the traits that enable them to successfully survive and reproduce.

Necrosis: Death of plant cells or tissues, leading to discoloration of leaves and stems.

Nitrogen Fixation: The process by which plants convert atmospheric nitrogen into nitrate or ammonium.

Outcrossing: The pollination which takes place between two different flowers that may or may not be the same cultivar.

Peduncle: The stem that supports either multiple flowers or a single flower.

Perennial: Plants which have a life cycle that lasts over two years.

Phloem: The vascular tissues that conduct food (nutrients) throughout the plant.

Pistil: The female organ of a flower which bears ovules or seeds – consisting of a complete ovary, style, and stigma.

Pollination: The process by which pollen is conveyed from the anther to the stigma (fertilization).

Protein: Plant tissue rich in organic molecules that provide necessary amino acids for human and/or livestock consumption.

Respiration: The chemical process whereby plants take in carbon dioxide and release oxygen to the atmosphere.

Rhizome: An underground root-like stem that has nodes and internodes.

Species: Refers to a group of plants that are closely related and whose members can interbreed.

Stamen: The male organ of the flower, consisting of the anthers and the filaments holding the anthers in place.

Stigma: The stigma is located at the apex of the style and is where pollen enters the pistil.

Stoma: An epidermal pore in leaf or stem of a plant that allows respiration and transpiration.

Style: The style is a narrow elongated part of the pistil between the ovary and stigma.

Tendril: A narrow stem-like appendage which wraps around surrounding plants to gain support.

Whorled: Arrangement of three or more leaves or other plant structures positioned at a node.

Xylem: Vascular tissue that conveys water and minerals throughout the plant.

21419473R00090

Made in the USA
Charleston, SC
18 August 2013